Tributes
Volume 43

Judgements and Truth
Essays in Honour of Jan Woleński

Volume 33
Logic and Computation. Essays in Honour of Amílcar Sernadas
Carlos Caleiro, Fransciso Dionísio, Paula Gouveia, Paulo Mateus and João Rasga, eds.

Volume 34
Models: Concepts, Theory, Logic, Reasoning, and Semantics. Essays Dedicated to Klaus-Dieter Schewe on the Occasion of his 60th Birthday
Atif Mashkoor, Qing Wang and Bernhrd Thalheim, eds.

Volume 35
Language, Evolution and Mind. Essays in Honour of Anne Reboul
Pierre Saint-Germier, ed.

Volume 36
Logic, Philosophy of Mathematics and their History.
Essays in Honor of W. W. Tait
Erich H. Reck, ed.

Volume 37
Argumentation-based Proofs of Endearment. Essays in Honor of Guillermo R. Simari on the Occasion of his 70th Birthday
Carlos I. Chesñevar, Marcelo A. Falappa, Eduardo Fermé, Alejandro J. García, Ana G. Maguitman, Diego C. Martínez, Maria Vanina Martinez, Ricardo O. Rodríguez, an Gerardo I. Simari, eds.

Volume 38
Logic, Intelligence and Artifices. Tributes to Tarcísio H. C. Pequeno
Jean-Yves Béziau, Francicleber Ferreira, Ana Teresa Martins and
Marcelino Pequeno, eds.

Volume 39
Word Recognition, Morphology and Lexical Reading. Essays in Honour of Cristina Burani
Simone Sulpizio, Laura Barca, Silvia Primativo and Lisa S. Arduino, eds

Volume 40
Natural Arguments. A Tribute to John Woods
Dov Gabbay, Lorenzo Magnani, Woosuk Park and Ahti-Veikko Pietarinen, eds.

Volume 41
On Kreisel's Interests. On the Foundations of Logic and Mathematics
Paul Weingartner and Hans-Peter Leeb, eds.

Volume 42
Abstract Consequence and Logics. Essays in Honor of Edelcio G. de Souza
Alexandre Costa-Liete, ed.

Volume 43
Judgements and Truth. Essays in Honour of Jan Woleński
Andrew Schumann, ed.

Tributes Series Editor
Dov Gabbay   dov.gabbay@kcl.ac.uk

Judgements and Truth
Essays in Honour of Jan Woleński

edited by

Andrew Schumann

© Individual authors and College Publications 2020. All rights reserved.

ISBN 978-1-84890-349-4

College Publications
Scientific Director: Dov Gabbay
Managing Director: Jane Spurr

http://www.collegepublications.co.uk

Cover design by Laraine Welch

All rights reserved. No part of this publication may be reproduced, stored in a retrieval system or transmitted in any form, or by any means, electronic, mechanical, photocopying, recording or otherwise without prior permission, in writing, from the publisher.

Contents

Andrew Schumann

Introduction 1

Roman Murawski

Proof vs Truth in Mathematics..13

Jean-Yves Beziau

The Mystery of the Fifth Logical Notion (Alice in the Wonderful Land of Logical Notions)...27

Kazimierz Trzęsicki

Idea of Artificial Intelligence...53

Alexandre Costa-Leite & Edelcio G. de Souza

Conjunctive and Disjunctive Limits: Abstract Logics and Modal Operators..99

Fabien Schang

A Judgmental Reconstruction of Some of Professor Woleński's Logical and Philosophical Writings......................................109

Jens Lemanski & Michał Dobrzański

Reism, Concretism and Schopenhauer Diagrams....................163

Tomasz Jarmużek & Mateusz Klonowski & Rafał Palczewski

Deontic Relationship in the Context of Jan Woleński's Metaethical Naturalism..191

Jerzy Pogonowski

A Note on Intended and Standard Models 211

Janusz Kaczmarek

About Some New Methods of Analytical Philosophy. Formalization, De-formalization and Topological Hermeneutics 227

Stanisław Krajewski

Anti-foundationalist Philosophy of Mathematics and Mathematical Proofs 251

Marcin Trepczyński

Necessity and Determinism in Robert Grosseteste's De libero arbitrio 269

Wojciech Krysztofiak

Logical Consequence Operators and Etatism 289

Marek Zirk-Sadowski

The Normative Permission and Legal Utterances 319

Introduction

Andrew Schumann

University of Information Technology and
Management in Rzeszow,
Sucharskiego 2
35-225 Rzeszow
Poland

e-mail: aschumann@wsiz.edu.pl

"Abraham begat Isaac; and Isaac begat Jacob; and Jacob begat Judas and his brethren; and Judas begat Phares and Zara of Thamar; and Phares begat Esrom; and Esrom begat Aram…" (*Matthew* 1:2–3). It is the beginning of the Gospel according to Matthew. It is a known sample of sacral genealogy in Christianity. Georg Wilhelm Friedrich Hegel (1770 – 1831) showed that philosophical ideas have their own genealogy, too. Moreover, each actual philosophical idea is nothing more than its true genealogy in the retrospective view or its long history in the perspective view, i.e. each idea is a development and transition from the state *an sich* (in itself) to the state *für sich* (for itself) [2] and it can be revealed only genealogically from the end of transition process or historically from the beginning of transition process.

Later Paul-Michel Foucault (1926 – 1984) presented genealogy as necessary method of philosophical analysis as such. According to him, each cultural or social phenomenon can be philosophically investigated only through its genealogical reconstruction. He started to distinguish between the epistemological level of knowledge presenting what is now and the genealogical reconstruction of existences. The genealogical reconstruction was called by him the "archaeological level of knowledge". It is one of the core objectives of philosophy:

> (…) archaeology, addressing itself to the general space of knowledge, to its configurations, and to the mode of being

of the things that appear in it, defines systems of simultaneity, as well as the series of mutations necessary and sufficient to circumscribe the threshold of a new positivity [1, p. xxv].

A genealogical reconstruction of ideas or looking for an archaeological level of knowledge can be found in logic, too. It means that logical ideas might be explicated through their genealogical analysis as well. Each significant logical theorem has some preliminary steps established by some proved propositions and these propositions constitute an inner history of the given theorem. Furthermore, we can focus on some philosophical intuitions and metatheoretical frameworks needed for formulating and proving this theorem. They also are a part of genealogical reconstruction within this theorem. Hence, a thorough understanding of logical statements implies an archaeological level of logic.

Jan Woleński (also known as Jan Hertrich-Woleński) was born 21 September 1940, in the same year as my father. From 1958 to 1963 he studied law at the Jagiellonian University and then from 1960 to 1964 philosophy at the same university. From the outset, his interest to logic was accompanied by analyzing the archaeological level of Polish logical tradition. Perhaps, it can be explained by his first law background – he tried to understand a copyright status of logical ideas through a reconstruction of genealogical trees of logical statements and concepts. He assembles a unique home library of logical works all his life and he remembers the names of all Polish logicians in the history of Poland. He became the grand master in explicating the archaeological level of Polish logic.

In the beginning of 20th century, the tradition of Polish logic was accumulated by the Lviv-Warsaw School (its former name was the Lvov-Warsaw School, its current name in Polish: *Szkoła Lwowsko-Warszawska*). Its most famous members are presented by Kazimierz Ajdukiewicz, Tadeusz Kotarbiński, Stanisław Leśniewski, Jan Łukasiewicz, and Alfred Tarski. Woleński showed that the Lviv-Warsaw School was an analytical school similar to the Vienna Circle in many respects [17]. In numerous papers, he reconstructed the archaeological level of logic for this school [11], [12], [13]. In his edited volumes [8], [9], [21], he popularized the history of this school among logicians. And in his monographs [14], [16], he presented an exhaustive review of the school. It is worth noting that in his recent project '*Lexicon of Polish Logicians 1900 – 1939*' (*Leksykon polskich logików 1900-1939*) supported by the grant from the Ministry of Science and Higher Education of Poland (0411/NPRH7/H30/86/2019 on the day of 02.10.2019), he is going to give a complete genealogical analysis of Polish logical ideas from 1900 to 1939. It will be a wonderful pearl of his many-years efforts in studying the history of Polish logic.

Woleński proved that within the archaeological level of Polish logic, Alfred Tarski (1901 – 1983) [10] is the most important logician. His semantic theory of truth [7], [19], [20], on the one hand, was "inspired by the Aristotelian tradition in philosophy, as well as the non-constructive style of working on the foundations of mathematics that was prevailed in Poland" [17], i.e. this theory has a reach genealogy in fact, and, on the other hand, this theory has a reach history after Tarski, too – many logicians follow this approach until now. An appropriate genealogy and history, as well as a complete explication, of Tarskian epistemological ideas are given in the following fundamental book of Jan Woleński: [18].

Tarski paid attention that the concept of truth must be defined for a definite formalized language **L**, but the definition itself should be formulated in the metalanguage **ML** [17], [20]. In the meanwhile, the definition should be formally correct, materially adequate, and satisfy a maximality of the set of truths in a given language **L**:

A sentence A of a language **L** is true if and only if it is satisfied by all infinite sequences of objects taken from the universe of discourse [17].

The Tarskian semantic theory of truth is explicated by Woleński in many papers and books [15], [21], [22].

I have to confess that Woleński's approach to genealogical analysis of logic inspired me to formulate my own research program of archaeology of logic. In this program we focus on studies of the history of early symbolic logic and its origin. According to these studies, symbolic logic was established in Babylonia [3], [4], [5]. Then it was developed in two concurrent branches: (1) within the Aramaic-Hebrew culture continued by the Talmud and Talmudic *middot* (logical inference rules for the Talmudic hermeneutics); (2) within the Greek logic presenting the Aristotelian syllogistic and the Stoic propositional logic. Then the Stoic logic had many impacts on establishing Nyāya logic [6]. The point is that Nyāya appeared in Gandhāra in the 2nd century A.D. at the time of Kaṇiṣka the Great. At this time the political elite remained Hellenized and the Greek language was official for more than 400 years before.

In this volume, there are collected new research papers devoted to judgments and truth. These papers take measure of the scope and impact of Woleński's views on truth conceptions, and present new contributions to the field of philosophy and logic. In *Proof vs Truth in Mathematics*', by Roman Murawski, relations between proofs and truth are analyzed. In *The Mystery of the Fifth Logical Notion (Alice in the Wonderful Land of Logical Notions)*', Jean-Yves Beziau discusses a theory presented in a posthumous paper by Alfred Tarski entitled *'What are logical notions?'*. In *'Idea of Artificial Intelligence'*, Kazimierz Trzęsicki gets the trace back on the development of Lullus's art, *ars combinatoria*, i.e. the author demonstrates a genealogical analysis of abstract machines. The paper *'Conjunctive and Disjunctive Limits: Abstract Logics and Modal Operators'*, by Alexandre Costa-Leite and Edelcio G. de Souza, introduces two concepts: conjunctive and disjunctive limits, to formalize levels of modal operators. In *'A Judgmental Reconstruction of Some of Professor Wolenski's Logical and Philosophical Writings'*, Fabien Schang concentrates on the nature of truth-values and their multiple uses in philosophy to genealogically explicate different means of using truth concepts. In *'Reism, Concretism and Schopenhauer Diagrams'*, Jens Lemanski and Michał Dobrzański showed that, according to Kazimierz Ajdukiewicz and Jan Woleński, there are two dimensions with which the abstract expression of reism can be made concrete: the ontological dimension and the semantic dimension. In *'Deontic Relationship in the Context of Jan Woleński's Metaethical Naturalism'*, Tomasz Jarmużek, Mateusz Klonowski, and Rafał Palczewski indicate how Jan Woleński's non-linguistic concept of norm allows us to clarify the deontic relationship between sentences and the given normative system. In *'A Note on Intended and Standard Models'* Jerzy Pogonowski discusses some problems concerning intended, standard, and non-standard models of mathematical theories with Woleński's views on these issues. In *'About Some New Methods of Analytical Philosophy. Formalization, De-formalization and Topological Hermeneutics'*, Janusz Kaczmarek continues the characteristics of philosophical methods specific to analytical philosophy, which were and are important for Jan

Woleński. In '*Anti-foundationalist Philosophy of Mathematics and Mathematical Proofs*', Stanisław Krajewski shows some main features of real proofs, such as being convincing, understandable, and explanatory. In '*Necessity and Determinism in Robert Grosseteste's De libero arbitrio*' Marcin Trepczyński follows the genealogical approach of Woleński and demonstrates that Robert Grosseteste's theory is still relevant and useful in contemporary debates, as it can provide strong arguments and enrich discussions, thanks to the two-perspectives approach, which generates some positions on the spectrum of determinism and indeterminism. In '*Logical Consequence Operators and Etatism*', by Wojciech Krysztofiak, there is presented the theory of logical consequence operators indexed with taboo functions to describe logical inferences in the environment of forbidden sentences. In '*The Normative Permission and Legal Utterances*' Marek Zirk-Sadowski proves that rejecting the existence of permissive norms and limitation of norms to prohibitions and commands alone is possible only with reducing the idea of function.

Acknowledgements

I am thankful to Jan Woleński for our fruitful friendship and I wish him many congratulations on the occasion of the 80th anniversary of his birth.

This volume was first published as an issue of *Studia Humana* (9, 3/4, 2020).

References

1. Foucault, M. *The Order of Things. An archaeology of the human sciences*, Taylor and Francis e-Library, 2005.
2. Hegel, G. W. F. *Vorlesungen über die Geschichte der Philosophie*, Erster Teil, Berlin: CreateSpace Independent Publishing Platform, 2013.
3. Schumann, A. Did the Neo-Babylonians Construct a Symbolic Logic for Legal Proceedings? *Journal of Applied Logics* 6(1), 2019, pp. 31-81.
4. Schumann, A. Legal Argumentation in Mesopotamia since Ur III, *Journal of Argumentation in Context* 9(2), 2020, pp. 243-282.
5. Schumann, A. On the Babylonian Origin of Symbolic Logic, *Studia Humana* 6(2), 2017, pp. 126-154.

6. Schumann, A. On the Origin of Indian Logic from the Viewpoint of the Pāli Canon, *Logica Universalis* 13(3), 2019, pp. 347-393.
7. Tarski, A., J. Tarski, and J. Woleński. Some Current Problems in Metamathematics, *History and Philosophy of Logic* 16, 1995, pp. 159-168.
8. Woleński, J. (ed.). *Kotarbinski: Logic, Semantics and Ontology*, Dordrecht, Boston: Kluwer Academic Publishers, 1990.
9. Woleński, J. (ed.). *Philosophical Logic in Poland*, Dordrecht, Boston: Kluwer Academic Publishers, 1994.
10. Woleński, J. Alfred Tarski (1901-1983), In A. Garrido and U. Wybraniec-Skardowska (eds.), *The Lvov-Warsaw School. Past and Present*, Basel: Birkhäuser, 2018, pp. 361-371.
11. Woleński, J. *Essays in the History of Logic and Logical Philosophy*, Cracov: Jagiellonian University Press, 1999.
12. Woleński, J. *Essays on Logic and Its Applications in Philosophy*, Frankfurt am Main: Peter Lang GmbH, Internationaler Verlag der Wissenschaften, 2011.
13. Woleński, J. *Historico-Philosophical Essays*, vol. 1, Kraków: Copernicus Center Press, 2013.
14. Woleński, J. *L'école de Lvov-Varsovie: Philosophie et logique en Pologne (1895-1939)*, Paris: Vrin, 2012.
15. Woleński, J. *Logic and Its Philosophy*, Berlin: Peter Lang, 2018.
16. Woleński, J. *Logic and Philosophy in the Lvov-Warsaw School*, Dordrecht, Boston: Kluwer Academic Publishers, 1989.
17. Woleński, J. Lvov-Warsaw School, *Stanford Encyclopedia of Philosophy*, https://plato.stanford.edu/entries/lvov-warsaw/ , 2020.
18. Woleński, J. *Semantics and Truth*, Heidelberg: Springer Nature, 2019.
19. Woleński, J. Some Philosophical Aspects of Semantic Theory of Truth, In A. Garrido and U. Wybraniec-Skardowska (eds.), *The Lvov-Warsaw School. Past and Present*, Basel: Birkhäuser, 2018, pp. 373-389.
20. Woleński, J. The Semantic Theory of Truth, *The Internet Encyclopedia of Philosophy*, https://www.iep.utm.edu/s-truth/, 2020.
21. Woleński, J., E. Köhler (eds.). *Alfred Tarski and the Vienna circle: Austro-Polish Connections in Logical Empiricism*, Dordrecht, Boston: Kluwer Academic Publishers, 1999.

22. Woleński, J., I. Niiniluoto, M. Sintonen (eds.). *Handbook of Epistemology*, Dordrecht, Boston: Kluwer Academic Publishers, 2004.

Prof. Jan Woleński at the Award of the Foundation for Polish Science (2013),
© https://www.fnp.org.pl/

Prof. Jan Woleński at awarding the title of Doctor Honoris Causa of Lodz University (2020),
© https://www.uni.lodz.pl/

Prof. Jan Woleński meets Prof. Saul Kripke (2017),
© Jan Woleński

Prof. Jan Woleński visits the monument 'Broken Hearth' installed on a former Jewish cemetery in Minsk (Belarus) as a memorial tribute to the victims of Nazism who died in a ghetto during the World War II (2016),
© Andrew Schumann

Proof *vs* Truth in Mathematics

Roman Murawski

Adam Mickiewicz University
Uniwersytetu Poznańskiego 4 Street
61-614 Poznań, Poland
e-mail: rmur@amu.edu.pl

1. Introduction

Concepts of proof and truth play an important role in metamathematics, especially in the methodology and the foundations of mathematics. Proofs form the main method of justifying mathematical statements. Only statements that have been proved are treated as belonging to the corpus of mathematical knowledge. Proofs are used to convince the readers of the truth of presented theorems. But what is a proof? What does it mean in mathematics that a given statement is true? What is truth (in mathematics)?

In mathematical research practice proof is a sequence of arguments that should demonstrate the truth of the claim. Of course, particular arguments used in a proof depend on the situation, on the audience, on the type of a claim, etc. Hence the concept of proof has in fact a cultural, psychological and historical character. In practice mathematicians generally agree on whether a given argumentation is a proof. More difficult is the task of defining a proof as such. Beside the concept of proof used in research practice there is a concept of proof developed by logic. What are the relations between those two concepts? What roles do they play in mathematics?

On the other hand the concept of truth belongs to the fundamental concepts that have been considered in epistemology since ancient Greece.[1] There were many attempts to define this vague concept. The classical definition (attributed to Aristotle) says that a statement is true if and only if it agrees with the reality, or – as

Thomas Aquinas put it: "Veritas est adequatio intellectus et rei, secundum quod intellectus dicit esse quad est vel non esse quod non est" (*De veritate*, 1, 2).

But what does it mean that a mathematical statement (for example: "2 + 2 = 4") agrees with the reality? With what reality? One can answer: "With mathematical reality?" But what is mathematical reality? And we come here to one of the fundamental problems of the ontology of mathematics: where and how do mathematical objects exist? Is the mathematical universe a reality or an artifact?

2. Proof in Mathematics: Formal *vs* Informal

Mathematics was and still is developed in an informal way using intuition and heuristic reasonings – it is still developed in fact in the spirit of Euclid (or sometimes of Archimedes) in a *quasi*-axiomatic way. Moreover, informal reasonings appear not only in the context of discovery but also in the context of justification. Any correct methods are allowed to justify statements. Which methods are correct is decided in practice by the community of mathematicians. The ultimate aim of mathematics is "to provide correct proofs of true theorems" [2, p. 105]. In their research practice mathematicians usually do not distinguish concepts "true" and "provable" and often replace them by each other. Mathematicians used to say that a given theorem holds or that it is true and not that it is provable in such and such theory. It should be added that axioms of theories being developed are not always precisely formulated and admissible methods are not precisely described.[2]

Informal proofs used in mathematical research practice play various roles. One can distinguish among others the following roles (cf. [4], [7]):
    (1) verification,
    (2) explanation,
    (3) systematization,
    (4) discovery,
    (5) intellectual challenge,
    (6) communication,
    (7) justification of definitions.

The most important and familiar to mathematicians is the first role. In fact only verified statements can be accepted. On the other hand a proof should not only provide a verification of a theorem but it should also explain why does it hold. Therefore mathematicians are often not satisfied by a given proof but are looking for new proofs which would have more explanatory power. Note that a proof that verifies a theorem does not have to explain why it holds. It is also worth distinguishing between proofs that convince and proofs that explain. The former should show that a statement holds or is true and can be accepted, the latter – why it is so. Of course there are proofs that both convince and explain. The explanatory proof should give an insight in the matter whereas the convincing one should be concise or general. Another distinction that can be made is the distinction between explanation and understanding. In the research practice of mathematicians simplicity is often treated as a characteristic feature of understanding. Therefore, as G.-C. Rota writes: "[i]t is an article of faith among mathematicians that after a new theorem is discovered, other, simpler proof of it will be given until a definitive proof is found" [23, p. 192].

It is also worth quoting in this context Aschbacher who wrote:

> The first proof of a theorem is usually relatively complicated and unpleasant. But if the result is sufficiently important, new approaches replace and refine the original proof, usually by embedding it in a more sophisticated conceptual context, until the theorem eventually comes to be viewed as an obvious corollary of a larger theoretical construct. Thus proofs are a means for establishing what is real and what is not, but also a vehicle for arriving at a deeper understanding of mathematical reality [1, p. 2403].

As indicated above a concept of a "normal" proof used by mathematicians in their research practice (we called it "informal" proofs) is in fact vague and not precise. In the 19th century there appeared a new trend in the philosophy of mathematics and in the foundations of it whose aim was the clarification of basic mathematical concepts, especially those of analysis (cf. works by Cauchy, Weierstrass, Bolzano, Dedekind). One of the drivers of this trend was the discovery of antinomies in set theory (due among others to C. Burali-Forte, G. Cantor, B. Russell) and of semantical antinomies (among others by G. D. Berry and K. Grelling). All those facts forced the revision of fundametal concepts of metamathematics.

One of the formulated proposals was the programme of David Hilbert and the formalism based on it. Hilbert's main aim was to justify mathematics developed so far, in particular to show that mathematics using the concept of an actual infinity is consistent and secure. To achieve this aim Hilbert proposed to develop a new theory called proof theory (*Beweistheorie*). It should be a study of proofs in mathematics – however not of real proofs constructed by mathematicians but of formal proofs. The latter played a fundamental role in Hilbert's programme. Hilbert proposed to formalize all theories of the entirety of mathematics and to prove the consistency of them. Note that he did not want to replace the mathematics developed by mathematicians by formalized theories – the formalization was for him only a methodological tool that should enable the study of theories as such.

To formalize a theory one should first fix a symbolic formal language with formal rules of constructing formulas in it, then fix appropriate axioms expressed in this language as well as accepted rules of inference which again should have an entirely formal and syntactic character. A proof (exactly: formal proof) of a formula φ in such a theory is now a sequence of formulas $\varphi_1, \varphi_2, \ldots, \varphi_n$ such that the last member of the sequence is the formula φ and all members of it either belong to the set of presumed axioms or are consequences of previous members of the sequence according to one of the accepted rules of inference. Observe that this concept of a formal proof has a syntactic character and does not refer to any semantical notions such as meaning or interpretation.

Note that formalization is connected also with the idea of mathematical rigor. Detlefsen [6, p. viii] writes:

> [W]ith the vigorous development of techniques of *formalization* that has taken place in this [i.e., 20th century – my remark, R.M.] century, demands for rigor have increased to a point where it is now the reigning orthodoxy to require that, to be genuine, a proof must be formalizable. This emphasis on formalization is based on the belief that the only kinds of inferences ultimately to be admitted into mathematical reasoning are *logical* inferences [. . .].

Comparing the usual proofs of mathematical research practice (informal proofs) and formal proofs one can see that both types of proofs consist of steps of deduction. They differ by the properties of those steps. According to Hamami [10] one can distinguish here three types of differences: formality, generality and mechanicality. Informal inferences are meaning dependent, matter dependent and content dependent wheras formal inferences are meaning, matter and content independent. Hamami [10, p. 679] writes: "To say that logical inference is *formal* is to say that it is governed by rules of inference which only depend on the logical form of premises and conclusion, and not on their meaning, matter, or content."

Tarski [25, p. 187] said: "[T]he relation of following logically is completely independent of the sense of the extra-logical constants occurring in the sentences among which this relation obtains […]."

Informal inferences are non-general wheras formal ones are general. This means in particular that the former are topic-specific, subject matter dependent and domain dependent, and the latter are topic-neutral, subject matter and domain independent. Detlefsen [5, p. 350] wrote in connection with this:

> The mathematician's inferences stem from and reflect a knowledge of the local "architecture" (Poincaré's term) of the particular subject with which they are concerned, while those of the logician represent only a globally valid, topic-neutral (and, therefore, locally insensitive!) form of knowledge.

Hamimi [10] explains that the claim that logical inference is general means in particular that "it is governed by rules of inference that are *generally applicable*, i.e., that are applicable to propositions – premisses and conclusions – belonging to any and every topic, subject matter, or domain" [10, pp. 684-685].

The last difference between informal and formal proofs distinguished by Hamimi is the property of mechanicality: informal ones are non-mechanical and formal ones – mechanical. What does it mean is explained by the following quotations. Kreisel [16, p. 21] writes:

> Mathematical reasoning, except in the 'limiting' case of numerical computations, does not present itself to us as the

execution of mechanical rules [. . .] The connection between reliability and the possibility of mechanical checking is usually, and somewhat uncritically, taken for granted.

And Hamimi [10, p. 695] says: "To say that logical inference is *mechanical* is to say that it is governed by rules of inference that are mechanical."

One can distinguish here two senses in which logical rules of inference are mechanical: mechanical applicability and mechanical checkability.

Add at the end of this section that the concept of a formal proof enables us to study mathematical theories as theories, to investigate their properties, etc. It makes possible the entirety of metamathematics. However, the following question arises: what are the relations between formal and informal proofs. Recall that the first one is a practical notion of a semantical character, not having a precise definition. The latter is a theoretical concept of a syntactical character used in logical studies. Mathematicians are usually convinced that every "normal", i.e., informal mathematical proof can be transformed into a formalized one, however there are no general rules describing how this can and should be done. This thesis is sometimes called Hilbert's thesis. Barwise [3] wrote:[3] "[T]he informal notion of provable used in mathematics is made precise by the formal notion *provable in first-order logic*. Following a sug[g]estion of Martin Davis, we refer to this view as *Hilbert's Thesis*."

In fact a formalization of an informal proof requires often some original and not so obvious ideas.

3. Truth in Mathematics

We indicated above that "normal" mathematicians (i.e., mathematicians not being logicians or specialists in the foundations of mathematics) do not distinguish in their research practice between provability (in the broad sense) and truth. Moreover, those two concepts are usually identified in practice. This was done also by formalists.[4] Gödel wrote in a letter of 7th March 1968 to Hao Wang [cf. 29, p. 10]: "[...] formalists considered formal demonstrability to be an *analysis* of the concept of mathematical truth and, therefore were of course not in a position to *distinguish* the two."

Note that "mathematical truth" should be understood here in an intuitive way. Moreover, the informal concept of truth was not commonly accepted as a definite mathematical notion in Hilbert's and Gödel's time. There was also no definite distinction between syntax and semantics. This explains also, in some sense, why Hilbert preferred to deal in his metamathematics solely with forms of formulas, using only finitary reasonings which were considered to be secure – contrary to semantical reasonings which were non-finitary (sometimes called: infinitary) and consequently not secure.

The precise definition of truth was given by Tarski in his famous paper *Pojęcie prawdy w językach nauk dedukcyjnych* [24]. Referring to the classical Aristotle's definition he attempted to make more precise the concept of truth with respect to formalized languages. In such languages "the sense of every expression is unambiguously determined by its form" [27, p. 186].

Tarski defined the concept of truth by using the concept of satisfaction, more exactly, satisfaction of a formula on a valuation by a given interpretation of primitive notions of the considered language, hence in a given structure. His definition refers to the so called convention (T) according to which the statement "Snow is white" is true if and only if snow is white. In fact Tarski did not give a definition of truth but defined only the class of true sentences (of a given language).

Tarski's definition has an infinitary character – the infinity appears in the reference to infinite sequences of elements of the considered structure (valuations) as well as in the case of satisfaction of formulas with quantifiers. It does not go beyond the extensional adequacy and does not explain the essence of the truth and of being true. It relativizes also the concept of truth to a given structure or domain.

In the above mentioned paper [24] Tarski formulated also the theorem on the undefinability of truth. It says that the conccept of truth for given formalized language cannot be definied in this language itself – to do this more powerful means are necessary. In other words: the set of sentences true in a given structure is not definable in it (though in some cases it is definable with parameters). Tarski formulated this theorem as Theorem I, point (β) [cf. 26, p. 247]:[5] "[A]ssuming that the class of all provable sentences of the metatheory is consistent, it is impossible to construct an adequate definition of truth in the sense of convention T on the basis of the metatheory."

One of the consequences of Tarski's theorem is the fact that in order to construct truth theory, for example, for the language of the arithmetic of natural numbers (hence a theory of finite entities) one should apply more powerful means, in fact the infinity. In other words: the concept of an arithmetical truth is not arithmetically definable. Generally: semantics needs the infinity! It indicates also the gap between the syntactical concept of a (formal) proof and (formal) provability on the one side and the concept of truth. In fact, for example, the set of true arithmetical sentences is not definable in the language of arithmetics whereas the set of provable sentences (theorems) of arithmetic is arithmetically definable, even more: it is definable by a simple formula (more exactly: by a formula with one existential quantifier and logical connectives as well as eventually bounded quantifiers). Hence one can say that the concept of truth transcends all syntactical means.

The indicated difference between the (definablity of the concept of) provability and (the undefinability of the concept of) truth was the key reason for the famous incompleteness theorems proved by Gödel [8]. Gödel wrote on his discovery in a draft reply to a letter dated 27th May 1970 from Yossef Balas, then a student at the University of Northern Iowa [30, pp. 84-85] and indicated there that it was precisely his recognition of the contrast between the formal definability of provability and the formal undefinability of truth that led him to his discovery of incompleteness. The first incompleteness theorem implies that in every consistent theory containing the arithmetic of natural numbers there are undecidable (i.e. that can neither be proved nor disproved) statements φ such that one formula of the pair φ and *non*-φ is satisfied/true in the intended (standard) model of the theory. It shows that (formal) provability is not the same as truth! However both these concepts are connected by the completeness theorem stating that a statement φ is a theorem of a theory T if and only if φ is true in *every* model of T. And theories usually possess (infinitely) many various models – not only the intended one (called: standard). So we have that:

1. if a formula φ is provable in the theory T then it is true in every model of T, hence also in the intended model of T,

2. it is not true that for any formula φ: if φ is true in the intended (standard) model of T then it is provable in T.

Add that when "normal" mathematicians are saying that a given sentence φ is true then they have in mind that it is true in the intended (standard) model.

One should mention also another phenomenon. As indicated above the concept of truth/true sentence for a given language L is not definable in the language L itself. However partial concepts of truth for formulas of L are definable in L. More exactly: if one considers only formulas of L with a given maximal number of quantifiers (this is in fact a restiction of the complexity of a formula) then the concepts of satisfaction and truth for such formulas of a language L are definable in L. It can be proved that the definition of the satisfaction predicate for formulas with maximally k quantifiers is a formula with k quantifiers, i.e., a formula of the same degree of complexity. Details can be found in our monograph [18].

The concept of truth/true formula can be investigated also by mathematical, more exactly: by axiomatic-deductive methods. Conditions formulated in Tarski's definition of truth can be treated as axioms characterizing the predicate of being satisfied and true. Such an approach has been studied in detail for the case of arithmetic of natural numbers – cf. for example [17] and [21].

Results obtained by described investigations show that not for every model of arithmetic one can define a concept of satisfaction and truth on it having natural properties assumed and required by Tarski's definition. A necessary condition is here the property that the model should be recursively saturated.[6] Additional properties of a model must be assumed if one requires that the concept of truth upon a given model have some useful (and natural) properties like being full (i.e., deciding the truth of every formula on any valuation) or being inductive (this property means that the induction principle holds not only with respect to formulas of the language of arithmetic but also for an extended language augmented by the satisfaction/truth predicate).

It also turns out that if a concept of satisfaction and truth (called a satisfaction class[7]) for a given structure can be defined then it can be done in many mutually inconsistent ways, i.e., if there exists a satisfaction class on the model then there exist many such satisfaction classes. This shows that the axiomatic characterization of the concept of satisfaction and truth based on Tarski's definition is not complete and unique, that Tarski's conditions (treated as axioms) are too weak. This phenomenon can be removed by allowing more powerful – for example set-theoretical – means. All this shows the complexity of the concept of truth.

We indicated above the gap between provability and syntactical concepts on the one hand and satisfaction/truth and semantical concepts on the other. However it turns out that the concept of truth can be (in a certain sense) replaced by the concept of consistency (hence: a syntactical concept) in the so called ω-logic (it is a generalization of the usual classical logic obtained by admitting the so called ω-rule and reasonings of infinite length) and by the transfinite induction.[8] This confirms the thesis that semantical concepts such as satisfaction and truth require infinitary means. Such concepts can be expressed or replaced by richer syntactical ones, however, this requires the resignation from the requirement of being finitary, in particular from the natural requirement that a proof must have a finite length and can refer only to finitely many assumptions.

4. Conclusion

In research practice mathematicians do not fix and do not restrict allowed methods of proof – any correct method is practically allowed. A mathematician wants to know what properties the considered and investigated structure (intended structure/model, standard structure/model) has or whether a particular property is true/holds in this structure. She/he is not interested in the problem of whether this property can be deduced from a certain given and restricted set of axioms. Therefore, for example, a specialist in number theory who investigates the structure of the natural numbers (i.e., the structure $(N, S, +, \cdot, 0)$ where N is the set $\{0, 1, 2, 3, ...\}$, S denotes the successor function, $+$ and $\cdot$ denote, resp., addition and multiplication of natural numebrs and 0 denotes the distinguished element called "zero") is not working in the framework of a fixed axiomatized formal system of arithmetic but is using any correct mathematical methods in order to decide whether a considered property is true/holds in the investigated structure (in the intended, standard model of arithmetic of natural numbers). Consequently she/he does not hesitate to use even methods of complex analysis (as is done in the analytic numer theory) if only they can be useful in deciding the considered problem.

The informal and vague concept of proof used by mathematicians in their research practice can be made precise by the concept of formal proof. The latter makes possible exact metamathematical investigations of mathematical theories – more exactly of their formal counterparts (and not of real theories considered by "normal" mathematicians). However the formal concept of proof (with precisely described and restricted rules of inference) as well as the very concept of formalized theory based on it have some limitations indicated by Gödel's incompleteness theorems. On the other hand the precise concept of satisfaction and truth relativizes truth to a given structure/interpretation. The concept of formal proof is adequate with respect to *all* models of a considered theory (as the completeness theorem states) and not only to the truth in the intended/standard structure. All this implies that metamathematical studies of proofs, structures, theorems and theories are not exact counterparts of what mathematicians are really doing in their research practice, they are in fact idealizations of the real practice.

Let us finish our considerations by quating Alfred Tarski who in the paper "Truth and proof" wrote:

> Proof is still the only method used to ascertain the truth of sentences within any specific mathematical theory. […] The notion of a true sentence functions thus as an ideal limit which can never be reached but which we try to approximate by gradually widening the set of provable sentences. […] There is no conflict between the notions of truth and proof in the development of mathematics; the two notions are not at war but live in peaceful coexistence [27, p. 77].

References

1. Aschbacher, M. Highly complex proofs and implications of such proofs. *Philosophical Transactions of the Royal Society (A)* 363, 2005, pp. 2401–2406.
2. Avigad, J. Mathematical method and proof, *Synthese* 153, 2006, pp. 105–159.

3. Barwise, J. An introduction to first-order logic, In J. Barwise (ed.), *Handbook of Mathematical Logic*, Amsterdam: North-Holland, 1977, pp. 5–46.

4. CadwalladerOlsker, T. What do *we* mean by mathematical proof? *Journal of Humanistic Mathematics* 1, 2011, pp. 33–60.

5. Detlefsen, M. Poincaré against the logicians, *Synthese* 90 (3), 1992, pp. 349–378.

6. Detlefsen, M. (ed.). *Proof, Logic and Formalization*, London: Routledge, 1992,

7. De Villiers, M. D. *Rethinking Proof with the Geometer's Sketchpad*, Emeryville, CA: Key Curriculum Press, 1999.

8. Gödel, K. Über formal unentscheidbare Sätze der *Principia Mathematica* und verwandter Systeme. I, Monatshefte für Mathematik und Physik 38, 1931, pp. 173-198. Reprinted with English translation: On formally undecidable propositions of *Principia Mathematica* and related systems, In Gödel K. *Collected Works*, vol. I: *Publications 1929-1936*, S. Feferman, J. W. Dawson, Jr., S. C. Kleene, G. H. Moore, R. M. Solovay and J. van Heijenoort eds.), New York: Oxford University Press, and Oxford: Clarendon Press, pp. 144–195.

9. Gödel K. *Collected Works*, vol. I: *Publications 1929-1936*, S. Feferman, J. W. Dawson, Jr., S. C. Kleene, G. H. Moore, R. M. Solovay and J. van Heijenoort eds.), New York: Oxford University Press, and Oxford: Clarendon Press.

10. Hamami, Y. Mathematical inference and logical inference, *The Review of Symbolic Logic* 11 (4), 2019, pp. 665–704.

11. Kahle, R. Is there a "Hilbert thesis"? *Studia Logica* 107, 2019, pp. 145–165.

12. Kaye, R. *Models of Peano Arithmetic*, Oxford: Clarendon Press, 1991.

13. Kotlarski, H., and Z. Ratajczyk. Inductive full saisfaction classes, *Annals of Pure and Applied Logic* 47, 1990, pp. 199-223.

14. Kotlarski, H., and Z. Ratajczyk. More on induction in the language with a full satisfaction class, *Zeitschrift für Mathematische Logik und Grundlagen der Mathematik* 36, 1990, pp. 441-454.

15. Krajewski, S. Non-standard satisfaction classes, In W. Marek, M. Srebrny and A. Zarach (eds.), *Set Theory and Hierarchy Theory*, Proc. Bierutowice Conf. 1975, Lecture Notes in Mathematics 537, Berlin-Heidelberg-New York: Springer Verlag, 1976, pp. 121-144.

16. Kreisel, G. The formalist-positivist doctrine of mathematical precision in the light of experience, *L'Âge de la Science* 3, 1970, pp. 17–46.
17. Murawski R. Satisfaction classes – a survey, In R. Murawski and J. Pogonowski (eds.), *Euphony and Logos*, Amsterdam/Atlanta, GA: Edition Rodopi, 1997, pp. 259–281.
18. Murawski, R. *Recursive Functions and Metamathematics. Problems of Completeness and Decidability, Gödel's Theorems*, Dordrecht/Boston/London: Kluwer Academic Publishers, 1999.
19. Murawski, R. Truth vs. provability – philosophical and historical remarks. *Logic and Logical Philosophy* 10, 2002, pp. 93–117.
20. Murawski, R. On the distinction proof-truth in mathematics, In P. Gärdenfors et al. (eds.), *In the Scope of Logic, Methodology and Philosophy of Science*, Dordrecht–Boston–London: Kluwer Academic, 2002, pp. 287–303.
21. Murawski, R. Troubles with (the concept of) truth in mathematics, *Logic and Logical Philosophy* 15, 2006, pp. 285–303. Reprinted in: R. Murawski, *Lógos and Máthēma. Studies in the Philosophy of Mathematics and History of Logic*, Frankfurt am Main: Peter Lang International Verlag der Wissenschaften, 2011, pp. 187–201.
22. Murawski, R. Some historical, philosophical and methodological remarks on proof in mathematics, In D. Probst and P. Schuster (Eds.), *Concepts of Proof in Mathematics, Philosophy, and Computer Science*, Ontos Mathematical Logic, Berlin: Walter de Gruyter, 2016, pp. 251–268.
23. Rota, G.–C. The phenomenology of mathematical proof, *Synthese* 111, 1997, pp. 183–196.
24. Tarski, A. *Pojęcie prawdy w językach nauk dedukcyjnych*, Warszawa: Towarzystwo Naukowe Warszawskie, 1933, Wydział III Nauk Matematyczno–Fizycznych, vol. 34. Reprinted in: A. Tarski, *Pisma logiczno-filozoficzne*, vol. 1: *Prawda*, Warszawa: Wydawnictwo Naukowe PWN, 1995, pp. 131–172. English translation: The concept of truth in formalized languages, In A. Tarski, *Logic, Semantics, Metamathematics. Papers from 1923 to 1938*, Oxford: Clarendon Press, 1956, pp. 152–278 and in A. Tarski, *Logic, Semantics, Metamathematics. Papers from 1923 to 1938*, second edition edited and introduced by J. Corcoran, Indianapolis: Hackett Publishing Co., 1983, pp. 152–283.

25. Tarski, A. On the concept of following logically, *History and Philosophy of Logic* 23, 1936/2002, pp. 155–196.
26. Tarski, A. *Logic, Semantics, Metamathematics. Papers from 1923 to 1938*, Oxford: Clarendon Press, 1956.
27. Tarski, A. Truth and proof, *Scientific American* 220, No. 6, 1969, pp. 63–77.
28. Tarski, A. *Logic, Semantics, Metamathematics. Papers from 1923 to 1938*, second edition edited and introduced by J. Corcoran, Indianapolis: Hackett Publishing Co., 1983.
29. Wang, H. *From Mathematics to Philosophy*, London: Routledge and Kegan Paul, 1974.
30. Wang, H. *Reflections on Kurt Gödel*, Cambridge, Mass: The MIT Press, 1987.
31. Woleński, J. *Semantics and Truth*, Logic, Epistemology and the Unity of Science 45, Berlin: Springer Verlag, 2019.

Notes

1. For the development of the concept of truth see Woleński [31]
2. For more on proofs in mathematics and their role see, for example, Murawski [22].
3. Cf. Kahle [11].
4. For the development of the process of distinguishing concepts of provability and truth see, for example, Murawski [19] and [20].
5. Add that in the footnote Tarski explicitly states that his proof of this theorem uses Gödel's method of arithemtization of syntax and his method of diagonalization, however he stresses that he obtained his result independently.
6. For definition see for example Kaye [12].
7. The concept of a satisfaction class was introduced in Krajewski [15] and studied among others by Roman Kossak, Henryk Kotlarski, Stanisław Krajewski, Alistair Lachlan, Roman Murawski, Zygmunt Ratajczyk.
8. Cf. Kotlarski and Ratajczyk [13] as well as [14].

The Mystery of the Fifth Logical Notion (Alice in the Wonderful Land of Logical Notions)

Jean-Yves Beziau

University of Brazil
Largo de São Francisco de Paula 1
20051-070 Rio de Janeiro, RJ, Brazil
Brazilian Research Council
Brazilian Academy of Philosophy
e-mail: jyb@ufrj.br

0. An Original Idea not to be Found in Logical Textbooks

The present paper is based on a posthumous piece by Tarski entitled "What are logical notions?" [47]. Alfred Tarski (1901 – 1983) is the most prominent logician of the 20th century together with Kurt Gödel (1906 – 1978). Everyone interested in logic has heard of him.[1]

However, the theory of *logical notions* as presented here by Tarski is not something in the mainstream. This theory does not appear in any logical textbook! How to explain this paradox?

Tarski had a great many original ideas. Although he is very famous among philosophical logicians for his theory of truth, and among mathematical logicians for the development of model theory, many of his ideas and works are still not well-known.

The *Collected Papers* of Tarski (1921 – 1979), prepared by Steven Givant and Ralph McKenzie, were published in 1986 by Birkhäuser in four volumes of about 700 pages each. These volumes contain mostly photographic copies of the papers in the original language in which they were written: French, German, Polish, English, without translation and presentation.[2]

At the end of the 1920s, Tarski developed the theory of the consequence operator, and for many years this theory was hardly known outside of Poland. The idea of this theory appeared for the first time in a two-page paper published in French in Poland in 1929 [43]. It was translated into English by Robert Purdy and Jan Zygmunt only in 2012, and it was published with a presentation by Jan Zygmunt in the *Anthology of Universal Logic* [58].[3]

In addition to papers, Tarski also published some books. His famous *Introduction to Logic and to the Methodology of Deductive Sciences* [44], which was translated into many languages, can still be considered, after nearly one century, one of the best introductions to logic for teaching the subject. His last book was co-written with Steven Givant[4] and published after his death: *A formalization of set theory without variables* [49]. It is also outside the main stream of the present logical theories, and it is related to the work of Ernst Schröder (1841 – 1902).[5]

The expression "logical notions" is not standard. A more standard way of speaking would be "logical concepts". And if we have a look at a textbook of logic and/or an encyclopedia, we will find as basic "stuffs" related to logic, things like connectives, truth-tables, quantifiers, variables, constants, proof, inference, deduction, completeness, incompleteness...[6]

If you speak about "diversity", one will imagine you are talking about politics or biology, not about a logical notion. But in this 1986 paper Tarski considers "diversity" to be a fundamental logical notion. What kind of diversity is he talking about?

In the present paper we will investigate and clarify these logical notions. Our paper is written for a large audience and can be understood by people who have little or even no knowledge of logic, showing that it is possible to go directly to the heart of logic without much sophistry.

1. Logical Notions according to Tarski and Lindenbaum in the Perspective of a Childlike Methodology

In "What are logical notions?" Tarski proposes to define logical notions as those invariants under any one-to-one transformation, something he presents as a generalization of an idea of Felix Klein (1849 – 1925), connected to the so-called "Erlangen program".

Tarski presented two main lectures on this topic:

- May 16, 1966, at Bedford College, the University of London, UK.
- April 20, 1973, at the State University of New York at Buffalo, USA.[7]

The paper "What are logical notions?" is related to these talks and the final version was prepared by John Corcoran who attended the second talk. Tarski approved the paper but it was published only posthumously, in 1986 in the journal *History and Philosophy of Logic*.

Corcoran is a famous scholar who wrote the excellent introduction to the second edition of *Logic, Semantics, Metamathematics* (1983) [46], a selection of papers by Tarski from 1923 to 1938, translated into English by J. H. Woodger. Since its publication this Tarski 1986 paper has been cited in hundreds of scholarly works. Currently it is first on its journal's most-cited list. It has been reprinted in *The Limits of Logic*, edited by S. Shapiro [41].

Alfred Teitelbaum and Adolf Lindenbaum

As Tarski himself says in this paper, the idea of characterizing logical notions in such a way already appears in a paper by Lindenbaum and himself in 1934 [35]. Adolf Lindenbaum (1904 – 1941) was the main collaborator and friend of Tarski when he was in Poland, so it makes sense to use the expression "Tarski-Lindenbaum logical notions" (cf. also the expression "Tarski-Lindenbaum algebra").

One may dispute the order of the name. And there is a joke in Poland saying that all the main Tarski's theorems of this period are due to Lindenbaum. Considering that Tarski's original family name was "Teitelbaum", to avoid confusion, we could create the name

"A.Lindenteitelbaum" and attribute to the corresponding character the joint work, ideas and results, of these two famous logicians.

Lindenbaum-Tarski's original paper is technical but related to a particular context; on the other hand, Tarski's posthumous paper is general but rather informal. The full theory of logical notions has not yet been systematically developed, however some important advances have been made, in particular by Gila Sher [42], Vann McGee [37] and Denis Bonnay (Bonnay did a PhD on the topic [21], and see also his 2006 survey paper: "Logicality and Invariance" [20]). Solomon Feferman made some critical comments about Sher and McGee approaches in a paper dedicated to George Boolos entitled "Logic, Logics, and Logicism" [22]; moreover Luca Bellotti wrote an interesting study of Tarski 1986 paper simply called "Tarski on logical notions" [1].

The aim of our present paper is not to directly and explicitly develop such a theory, but to precisely analyze some aspects of it through a very simple case. Hopefully, this will contribute to the general theory. Right now there is a contrast between the fact that this 1986 Tarski paper is well- known among a small class of specialists but not among the wide class of people interested in logic, despite its profound interest.

We will focus here on a very simple case, logical notions in the context of binary relations (presented on page 150 of Tarski 1986 paper). We believe that the careful study of simple cases is an important task. Some people may avoid doing that thinking it is not serious, that it is trivial and childish. But as Alexander Grothendieck (1928 – 2014) wrote: "Discovery is the privilege of the child: the child who has no fear of being once again wrong, of looking like an idiot, of not being serious, of not doing things like everyone else."[8] And Adolf Lindenbaum himself was interested in the question of simplicity (cf. [34]).[9]

Many people are afraid of being too simple, or of expressing themselves in a too simple way. If you say something simple which is wrong, then you have more chance to be detected than if you were to say something wrong in a complicated way. If you don't speak clearly and someone says that what you are saying is wrong, you can always say the person made a wrong interpretation of what you wanted to say. A common trick among sophists. Simplicity is risky. But as they like to sing in Germany: *No Risk, No Fun!*

There are two complementary reasons to use a childlike methodology. On the one hand by doing that one may go to the root of things, if any. On the other hand, there is a pedagogical aspect: to explain the depth and interest of a topic to people having little knowledge of it. We would be delighted and it would be wonderful if a 7-year old girl like Alice could understand this paper. And we think it is possible.

There is a tendency to underestimate the intelligence of young children. But Patrick Suppes, with whom I was working for two years at Stanford at the very beginning of this century, brilliantly showed that a 7-year old can understand many things, through his EPGY program for young children, teaching them advanced mathematics, physics, music…

This does not mean that the present paper is restricted to children; we would be even more delighted if at the same time some adults enjoy the present paper and learn something, understand something. As written by Solomon in the *Proverbs* (3.13): "Joyful is the person who gains understanding."

2. The Four Tarski-Lindenbaum Logical Notions in the Case of a Binary Relation

We consider binary relations, i.e., relations between two objects, elements, things… There are many such relations and in fact, it is possible to prove that any *n*-ary relation can be expressed/reduced to a binary relation.[10] Tarski says the following about logical notions in case of binary relations:

> A simple argument shows that there are only four binary relations which are logical in this sense: the universal relation which always holds between any two objects, the empty relation which never holds, the identity relation which holds only between "two" objects when they are identical, and its opposite, the diversity relation. So the universal relation, the empty relation, identity, and diversity – these are the only logical binary relations between individuals. This is interesting because just these four relations were introduced and discussed in the theory of relations by Peirce, Schröder, and other logicians of the nineteenth century [47, p. 150].

Let us consider a binary relation on a set with two elements. The four relations can be represented by the following picture that is potatograph-like, popular in modern mathematics, and easy to understand for Alice (cf. [38], [39]). We have put the corresponding names below each one with the obvious corresponding substantive, but we have replaced "diversity" by "difference", because this is a better name. Hopefully Tarski will forgive us.

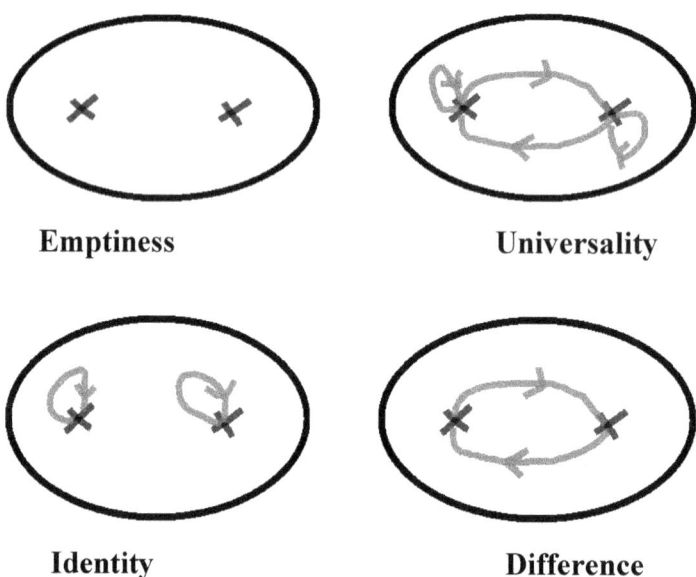

Emptiness **Universality**

Identity **Difference**

3. An Example of a Non-Logical Relation, Formulas and Models

Alice may ask: what does it mean that these and only these relations are logical? For example, why isn't the following one logical?

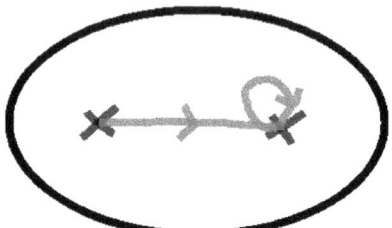

We say to Alice: try to describe this configuration (CONF1a) without giving a name to the two objects represented by the two crosses, and without referring directly to them. You cannot say, "The guy *on the*

left is not in relation with himself" nor "There is a guy who is in relation with *another* guy", but you can say "There is a guy who is in relation with himself" and "There is a guy who is in relation with a guy".

Alice may propose the following description: "There is someone who is not in relation with himself but who is in relation with someone in relation with himself (so the first someone cannot be the second someone), not in relation with him". It is correct, but this is not the only possible description.

This can be transcribed into the following formula ϕ:

$$\exists x \, (\neg(xRx) \, \land \, \exists y \, ((yRy) \land (xRy) \land \neg(yRx)))$$

This is a formula of first-order logic without equality (FoLoWoE). Alice may point out that this formula also describes the following configuration (CONF1b).

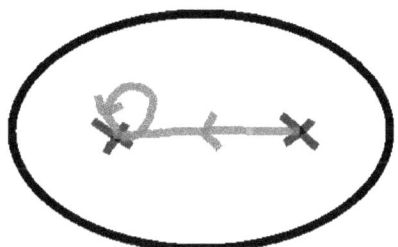

And she asks: is this not a problem?

To reply to this question, we have to introduce *model theory* to Alice, a theory developed by Alfred Tarski himself. Configurations described by a formula are called *models* of this formula. The notion of "model" in this sense was put forward by Tarski; he developed a whole theory explaining how this works [45].

Alice's question corresponds to the following two interrelated questions:
1) Is it a problem that our formula ϕ describing the first configuration also has a different configuration as a model?
2) Is it possible to find a first-order formula having as a model only the first configuration?

If we allow only formulas with no specific names, no constants, only variables, the answer to question (2) is negative. And this is not necessarily a problem because these two models are considered to be *isomorphic*: we can establish a one-to-one correspondence between the two that preserves the given structure of this configuration, which in model theory indeed is simply called a *structure*. This is because what is important is the structure, not the nature of individuals, who have no existence by themselves, outside a given structure.

The two crosses have been treated by Alice as if they were human beings by using the pronoun "someone". She could have said: "There is an object" or "There is something". But her choice is good because "someone" is a single word. "Something" also is single, but its meaning is not clear in the sense that "something" can refer to anything, like a storm, with many rain drops. This is not a good means to emphasize unicity, individuality. Tarski talks about *individuals*: "these are the only logical binary relations between individuals" [47, p. 150].

Furthermore, "someone" gives a lively touch to our discourse, one that is more amusing than disturbing. And something fundamental is preserved in this funny way of talking: anonymity. In French at some point in modern mathematics people were using expressions such as "truc", "machin", "bidule", a sense of surrealistic poetry that unfortunately has been lost.

Now Alice asks: why is CONF1a not a logical notion? We reply to her: consider a structure with three elements. Can you see that in this case the formula ϕ is not *categorical* in the sense that it has various non-isomorphic models: for example one model in which the additional third guy has no relations with the two others and one in which he is related with one of the two:

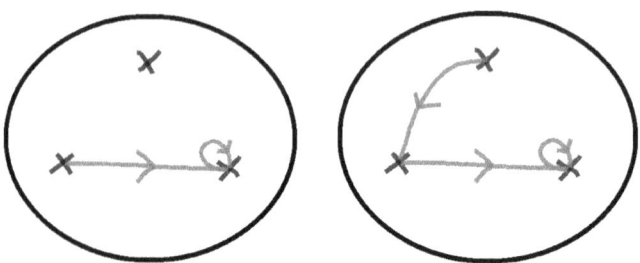

And that's the reason why:
- the formula ϕ does not describe a logical notion
- the relation in CONF1a is not considered as a logical notion.

Then Alice may inquire about these two reasons and their relations, asking:

(A1) As far as I understand, the formula ϕ does not describe a logical notion, because there is a cardinality for which it is not categorical, so categoricity is a necessary condition for logicality, but is it a sufficient reason? That is, if a formula ψ is categorical for each cardinality, does ψ describe a logical notion?

(A2) If a binary relation can be described by a categorical formula, is it sufficient to consider it to be a logical notion?

(A3) Is a binary relation considered to be a logical notion only if it can be described by a categorical formula?

The reply to (A1) and (A2) is positive because Tarski-Lindenbaum's logical notions are defined by *invariance*, expressed here by the notions of isomorphism and categoricity. The answer to question (A3) is not so obvious.

4. Expression and Formalization of the Four Tarski-Lindenbaum Logical Notions

Let us investigate with Alice the formulations of the four logical notions. We first point out to Alice that, "There is someone which is not in relation with himself but who is in relation with someone in relation with himself, not in relation with him" is rather complicated. And ask her to compare with the following formulations of the four logical notions:

Names	Formulations in Natural Language
Emptiness	Nobody is in relation with anybody
Universality	Everybody is in relation with everybody
Identity	Everybody is in relation only with himself
Difference	Everybody is in relation with everybody except with himself

The four relations have been expressed in this table using English, a natural language which spontaneously grew in the beautiful island where Alice was born. Now let us see how these four relations can be formulated in the artificial symbolic language FoLoWoE that we already presented to Alice in the previous section. Alice may draw the following table:

Names	Formulas of First-Order Logic without Equality
Emptiness	$\forall x \forall y \ \neg(xRy)$
Universality	$\forall x \forall y \ (xRy)$
Identity	???
Difference	???

She put some question marks where she was not able to find a formalization using FoLoWoE. There are in fact no formulas of FoLoWoE that express the logical notions of identity and difference. It has been proven that identity cannot be expressed in first-order logic without equality (see [2], [4], [5], [7], [9], [30]). We will not present the proof here, because this can be understood only after a full year's introductory class in logic (and some people have studied logic for one thousand and one nights and still don't understand that).

But admitting this theorem, Alice can immediately understand that the difference also cannot be expressed with a FoLoWoE formula, because, if it were the case, then the negation of if would express identity. All this gives a negative answer to the third Alice's question (A3).

Alice then may ask: but how do we know that identity and difference are logical notions? We can reply to her: close your eyes and imagine a structure with 5 elements where the only arrows you have are 5 arrows rounding above each of the five crosses, a generalization of the diagram we presented previously in the case of a structure with two elements. Does not this correspond to the expression, "Everybody is in relation only with himself", in the case of a 5-element set? Can you see something else corresponding to this expression in this case? And Alice of course after opening her eyes cannot reply no. We may go further and ask her to close her eyes again and imagine a similar structure with an infinite number of crosses, and she will certainly again not reply no.

The situation of difference is more difficult to imagine as a mental image, but we can ask Alice to draw a picture:

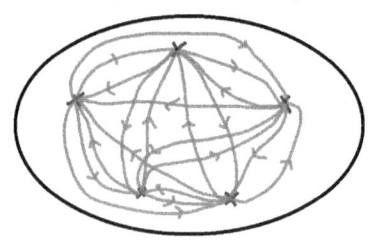

And this is the only configuration corresponding to difference in the case of a 5-element set that she can draw.

So, the situation of identity and difference is the same as the situation of universality and emptiness: they are categorical notions. But in the case of universality and emptiness this categoricity can be expressed by FoLoWoE formulas.

Alice may inquire why we forbid the use of the equality sign, "=", which is such a nice sign, invented by her cousin Robert Recorde! And she might argue that, if we lift the ban, she can express identity with the following formula:

$$\forall x \ (xRx) \land \forall y \ (\neg(y = x) \to \neg(xRy) \land \neg(yRx))$$

But we can say to Alice: is it not a vicious circle to define identity using equality, and is the equality sign not referring to identity? After thinking for half a second, she replies: "Sure and I don't want to be trapped in a vicious circle, long live freedom!" (cf. [17]).

5. Relations Between the Four Tarski-Lindenbaum Logical Notions

Now Alice may ask: what are the relations between these four logical notions? Tarski says that the relation of difference (that he calls "diversity") is the "opposite" of the relation of identity.

According to the theory of *the square of opposition*, there are three different notions of opposition: *contrariety*, *subcontrariety* and *contradiction*. In set theory, the notions corresponding to these three oppositions are respectively, *mutual exclusion* (or disjointness), *full intersecting union*, and *complementation*. Only the last word is standard.

Anyway, here are some diagrams corresponding to these notions, so that Alice will perfectly understand the meaning of these words:

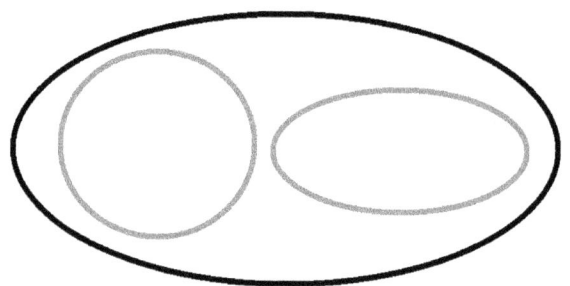

Mutual Exclusion

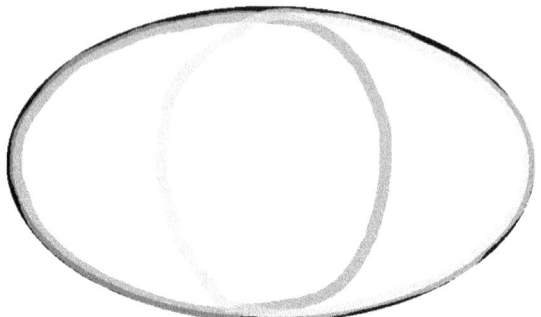

Full Intersecting Union

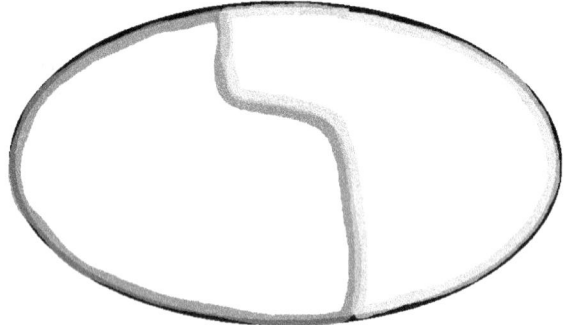

Complementation

A binary relation over a set of two distinct elements, glamorously called *"a"* and *"b"*, can be represented by a set of pairs. There are four possible pairs: <*a;a*>, <*a;b*>, <*b;a*>, <*b;b*>. The binary relation acting on them gives rise to the table below, also corresponding to what is called a Robinson's diagram – in honor of Abraham Robinson (1918 – 1974), a good friend of Tarski and also a great model-theorist.

Identity	Difference	Universality	Emptiness
(aRa)	$\neg(aRa)$	(aRa)	$\neg(aRa)$
$\neg(aRb)$	(aRb)	(aRb)	$\neg(aRb)$
$\neg(bRa)$	(bRa)	(bRa)	$\neg(bRa)$
(bRb)	$\neg(bRb)$	(bRb)	$\neg(bRb)$

This means, in the case of the relation of identity, that this relation is the set with the only two pairs: $<a;a>$, $<b;b>$, and in the case of the relation of difference that it is the set with only the two pairs: $<a;b>$, $<b;a>$. So, from the point of view of the set of all pairs, identity is the complement of difference, and vice-versa. For this reason, we can say that these two logical notions are in contradictory opposition, or, simply are contradictory. And the same happens between universality and emptiness: these two logical notions are contradictory. We can therefore draw the following healthy red cross picture:

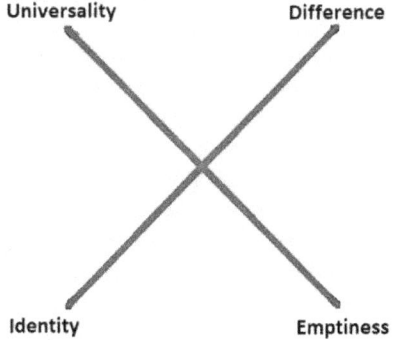

This red cross is a step towards a full square of opposition, where, besides contradiction in red, we have contrariety in blue, subcontrariety in green,[11] and in black subalternation (which is not an opposition), as shown in the figure below, where at each corner we have put quantifiers, having then the most typical exemplification of the square.

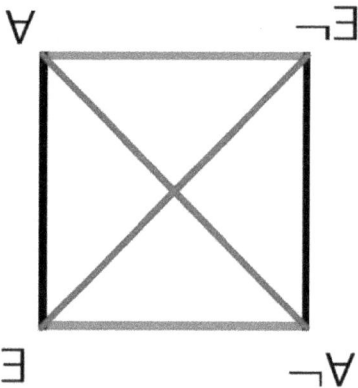

Alice may ask: can we make such a square of opposition with these four logical notions? The reply is negative. The fact that universality as a logical notion is expressed by a formula using universal quantifiers $\forall x \forall y\, (xRy)$ can be misleading, giving the idea that we can easily build a square of logical notions starting with the top left corner. But Alice can check that the relations between the four logical notions are properly described as follows:[12]

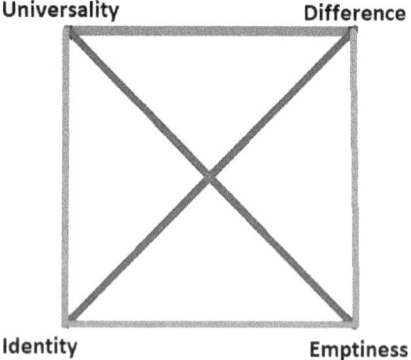

6. The Logicality of Variety

Besides the four structures corresponding to the four logical notions, there are in the simple case of a binary relation 12 other structures. This is just the world of combinatorics: we have a total of 16 structures for all the configurations of a binary relation over a two-element set. Among these 12 non-logical structures, half of them are reverse isomorphic images of the other ones – mirrors of them. In section 3, we have already presented two of them; here is the whole picture for Alice:

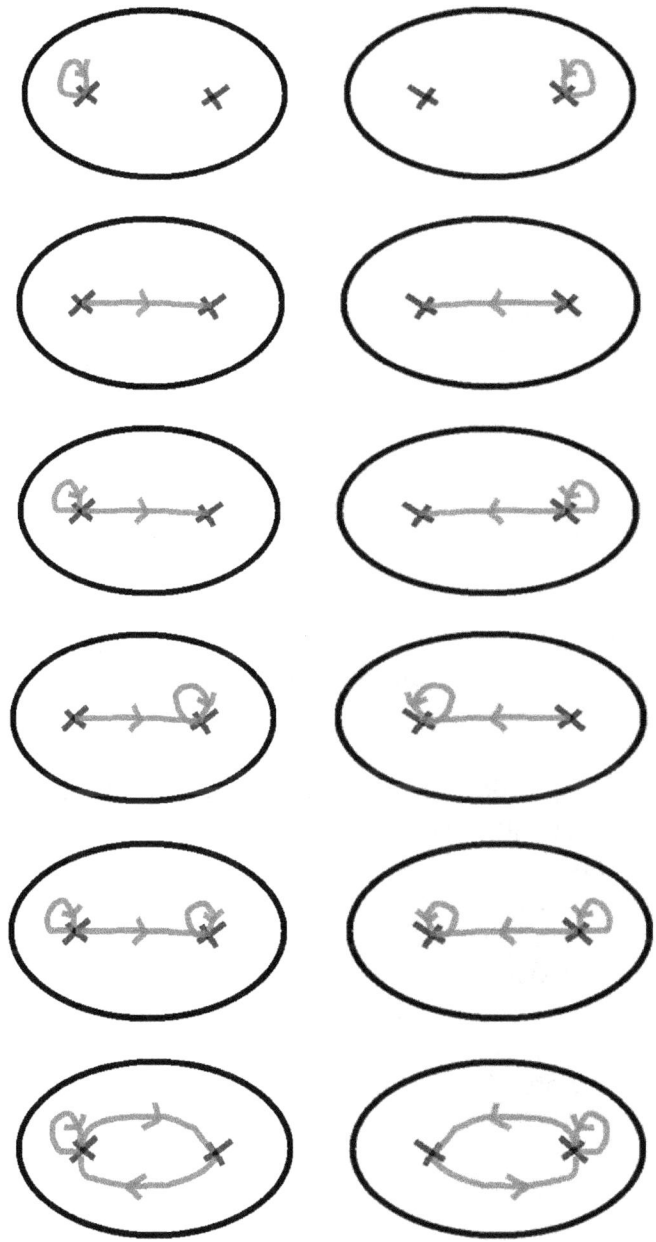

Let us consider the class of these 12 structures. It is the complement of the class of the 4 structures corresponding to logical notions. In this class of 12 structures there are non-isomorphic structures, for example:

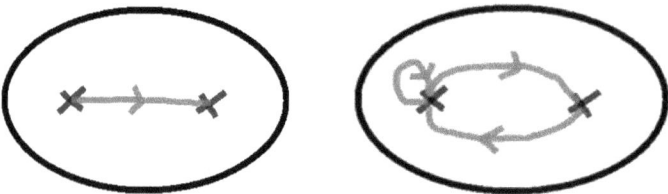

and Alice can easily be convinced that it will always be the case also for other cardinalities greater than 2. For this reason, we will say that this class corresponds to a notion, that we call *variety*.[13]

There is invariance in this variety: for every cardinality, it always refers to the same class of models, those not corresponding to logical notions. Alice may want to qualify variety as a non-logical notion. And, indeed, the notion of variety collects all the non-logical relations. But since it is invariant, and since invariance is the basis of Tarski-Lindenbaum logical notions, why not also saying that variety is a logical notion, a fifth logical notion? Tarski-Lindenbaum invariance is based on isomorphism, but it can be seen from the higher perspective of notions always referring to the same classes of models.

From the point of view of classes of models, the notion of variety is the contradictory opposite of logical relations, but this is not necessarily a problem, an obstacle to calling it a logical notion; contradictory opposition is a logical concept and we can apply here the idea of the identity of opposites.

In a previous paper [14] we were not afraid to claim that anticlassical logic, i.e. the complement of the consequence relation of classical logic, can be considered as a logic, even if it is obeying none of the three Tarskian axioms for a consequence relation (reflexivity, monotonicity and transitivity). We did that with the benediction of Jan Łukasiewicz who promoted the notion of a refutation system.

Here we are claiming that variety is a logical notion with the benediction of Alice Lindenteitelbaum.

7. An Enigma for Alice

For a happy ending we ask Alice: is there a FoLoWoE formula λ whose models are exactly the variety of non-logical relations (for any cardinality)?

Alice may propose the following formula λ:

$$\exists x \exists y \, (xRy) \land \exists x \exists y \, \neg(xRy) \land \exists x \neg(xRx) \land \exists x \, (xRx)$$

having in mind the table below where each negation of a logical notion is formulated by a FoLoWoE formula:

Name	Formulas of First-Order Logic without Equality
Non-Emptiness	$\exists x \exists y \, (xRy)$
Non-Universality	$\exists x \exists y \, \neg(xRy)$
Non-Identity	$\exists x \, \neg(xRx)$
Non-Difference	$\exists x \, (xRx)$

But this is a wrong answer! Because λ excludes the structures on lines 2 and 5 presented in the whole picture of non-logical relations in section 6. So we will let Alice find the answer to this question before the end of the night or before the end of her life…. . If she cannot find the answer by herself, we let her use as a joker MIAOU, the white cat, to whom she may ask the question (she can also have a look under the carpet):

8. Dedication and Personal Recollections

When X writes a paper in honor of Y, there are three exclusive and exhaustive categories forming a triangle of contrariety. X may write something which is:
(1) a critical comment of some work of Y
(2) related to the work of Y
(3) on a topic upon which X is working, but not in the two above categories.

The present paper clearly falls in the second category, for two reasons:
- The Polish School
- The Square of Opposition

Jan Woleński is mainly known for all the work he did to preserve and promote the history of the Lvov-Warsaw school of logic.[14] But he has also developed research in many topics, including the square of opposition.

We have never worked directly together, but we have collaborated in many projects. As far as I remember, my first encounter with Woleński was at the *38th Conference of History of Logic*, November 17-18, 1992, in Kraków, Poland and the latest one at the *41st International Wittgenstein Symposium*, August 5-11, 2018, in Kirchberg, Austria of which we both were invited speakers. In between we met in many other events around the world such as *Logic, Ontology, Aesthetics - The Golden Age of Polish Philosophy*, September 23-26, 2004, organized by Sandra Lapointe in Montreal, Canada. It would be difficult to list them all. What is important to stress is that this shows that both of us think that participation in events and interaction with colleagues are fundamental to research. Woleński also organized events. I remember in particular the *11th International Congress of Logic, Methodology and Philosophy of Science*, August 20-26, 1999, Kraków, Poland, the best LMPS I took part in.

I have also organized many events, in particular, launching three series of world events:
- UNILOG: *World Congress and School on Universal Logic*
- SQUARE: *World Congress on the Square of Opposition*
- WoCoLoR: *World Congress on Logic and Religion*[15]

Woleński has been an invited speaker of editions of all these series.[16] He was keynote at the 1st SQUARE in Montreux,

Switzerland, 2007, keynote at the 2nd WoCoLoR in Warsaw, Poland, 2017 (logically supporting atheism), keynote at the 2nd UNILOG in Xi'an, China, 2007.

At this event in China I also invited his former teacher Stan Surma whom he had not seen for many years (Surma emigrated during the communist period to Africa, then Australia, then New Zealand). In the photo in the next page you can see Jan Woleński circled in red, Stan Surma in green and me in blue. And you can also recognize other famous logicians such as Wilfrid Hodges, Arnon Avron, Bob Meyer, Vincent Hendricks, Arnold Koslow, Peter Schroeder-Heister, Valentin Goranko, Heinrich Wansing, etc.

Besides events, we have been collaborating in editorial projects. Jan Woleński wrote two entries for the *Internet Encyclopedia of Philosophy* of which I am logic area editor:
- Adolf Lindenbaum [56]
- The Semantic Theory of Truth [57]

He contributed to the volume *The Lvov-Warsaw School. Past and Present* edited by Á.Garrido and U.Wybraniec-Skardowska (2018) that I supervised as the managing editor of the book series *Studies in Universal Logic* where it was published. He wrote the following three chapters in this book:
- Alfred Tarski (1901 – 1983) [53]
- Some Philosophical Aspects of Semantic Theory of Truth [54]
- Jerzy Słupecki (1904 – 1987) [55][17]

He also published a paper on the square of opposition in the journal *Logica Universalis* that I founded and of which I am the Editor-in-Chief:
- Applications of squares of oppositions and their generalizations in philosophical analysis (2008) [52].

For all these reasons I am very glad to contribute to this special issue and to dedicate the present paper to Jan Woleński for his 80th birthday:

May you live actively to 120 years of age at least, Jan!

Acknowledgements

Thanks to Andrew Schumann for inviting me to contribute to this special issue. A previous version of this paper has been improved by information and comments provided by many friends : Arnon Avron, John Corcoran, Brad Dowden, Mike Dunn, Melvin Fitting, Rodrigo Freire, David Fuenmayor, Katarzyna Gan-Krzywoszynska, Val Goranko, Brice Halimi, Lloyd Humberstone, Srećko Kovač, Décio Krause, Arnold Koslow, Laurent Lafforgue, Dominique Luzeaux, David Makinson, David W.Miller, Daniele Mundici, Francesco Paoli, Daniel Parrochia, Rohit Parikh, Anca Pascu, Arnaud Plagnol, Robert Purdy, Stephen Read, Christophe Rey, Pascale Roure, Jean Sallantin, Sergey Sudoplatov, Alasdair Urquhart, Ioannis Vandoulakis, Denis Vernant, Jorge Petrucio Viana. I am grateful to all of them as well as to Alice and to my cat Miaou who both kindly took part to the project.

References

1. Bellotti, L. Tarski on logical notions, *Synthese* 135, 2003, pp. 401-413.
2. Beziau, J.-Y. Identity, logic and structure, *Bulletin of the Section of Logic* 25, 1996, pp. 89-94.
3. Beziau, J.-Y. New light on the square of oppositions and its nameless corner, *Logical Investigations* 10, 2003, pp. 218-232.
4. Beziau, J.-Y. Quine on identity, *Principia* 7, 2003, pp. 1-5.

5. Beziau, J.-Y. What is the principle of identity? (identity, logic and congruence), In F. T. Sautter and H. de Araújo Feitosa (eds), *Logica: teoria, aplicaçõoes e reflexões*, Campinas: CLE, 2004, pp. 163-172.
6. Beziau, J.-Y. Les axiomes de Tarski, In R. Pouivet and M.Rebuschi (eds), *La philosophie en Pologne 1918-1939*, Paris: Vrin, 2006, pp. 135-149.
7. Beziau, J.-Y. Mystérieuse identité, In *Le même et l'autre, identité et différence - Actes du XXXIe Congrès International de l'ASPLF*, Budapest: Eotvos, 2009, pp. 159-162.
8. Beziau, J.-Y (ed.). *Universal Logic: An Anthology*, Basel: Birkhäuser, 2012.
9. Beziau, J.-Y. Identification of identity, special Dale Jacquette memorial issue of *IfCoLog Journal of Logics and their Applications*, J. Woods (ed.), 4, 2017, pp. 3571-3581.
10. Beziau, J.-Y. The Pyramid of Meaning, In J. Ceuppens, H. Smessaert, J. van Craenenbroeck and G. Vanden Wyngaerd (eds.), *A Coat of Many Colours - D60,* Brussels, 2018.
11. Beziau, J.-Y. Logic Prizes *et Cætera*, *Logica Universalis* 12, 2018, pp. 271-296.
12. Beziau, J.-Y. The Lvov-Warsaw School: A True Mythology, In A. Garrido and U. Wybraniec-Skardowska (eds), *The Lvov-Warsaw School. Past and Present*, Basel: Birkhäuser, 2018, pp. 779-815.
13. Beziau, J.-Y. 1st World Logic Day: 14 January 2019, *Logica Universalis* 13, 2019, pp. 1-20.
14. Beziau, J.-Y. and Buchsbaum, A. Let us be Antilogical: Anti-Classical Logic as a Logic, In A. Moktefi, A. Moretti and F. Schang (eds.), *Soyons logiques / Let us be Logical*, London: College Publications, 2016, pp. 1-10.
15. Beziau, J.-Y., and J. Lemanski. The Cretan Square, *Logica Universalis* 14, 2020, pp. 1-5.
16. Beziau, J.-Y., V. Vandoulakis. *The Exoteric Square of Opposition*, Basel: Birkhäuser, 2020.
17. Beziau, J.-Y. Identity and equality in logic, mathematics and politics, In J.-Y. Beziau, J.-P-Desclés, A. Moktefi and A. Pascu (eds), *Logic in Question*, Basel: Birkhäuser, 2020.
18. Beziau, J.-Y, and S. Read (eds). Special issue of *History and Philosophy of Logic on the Square of Opposition* 35, 2014.
19. Birkhoff, G. On the structure of abstract algebras, *Proceedings of the Cambridge Philosophical Society* 31, 1935, pp. 433-454.

20. Bonnay, D. Logicality and Invariance, *Bulletin of Symbolic Logic* 14, 2006, pp. 29-68.
21. Bonnay, D. *Qu'est-ce qu'une constante logique?*, Ph.D. Dissertation, University Panthéon-Sorbone, Paris 1, 2006.
22. Feferman, S. Logic, Logics, and Logicism, *Notre Dame Journal of Formal Logic* 40, 1999, pp. 31-54.
23. Feferman, A. B., and S. Feferman. *Alfred Tarski: Life and Logic*, Cambridge: Cambridge University Press, 2004.
24. Corcoran, J. Categoricity, *History and Philosophy of Logic* 1, 1980, pp. 187- 207.
25. Corcoran, J. Tarski on logical notions (abstract), *Journal of Symbolic Logic* 53, 1988, p.1291.
26. Garrido, Á., and U. Wybraniec-Skardowska (eds). *The Lvov-Warsaw School. Past and Present*, Basel: Birkhäuser, 2018.
27. Givant, S. A portrait of Alfred Tarski, *The Mathematical Intelligencer* 13, 1991, pp. 16-32.
28. Givant, S. Unifying threads in Alfred Tarski's work, *The Mathematical Intelligencer* 21, 1999, pp. 47-58.
29. Grothendieck, A. *Récoltes et semailles - Réflexions et témoignage sur un passé de mathématicien*, unpublished manuscript, 1983-1986.
30. Hodges, W. Elementary predicate logic, In D. Gabbay and F. Guenthner (eds.), *Handbook of Philosophical logic*, vol. I, Dordrecht: Reidel, 1983, pp. 1-131.
31. Kalmar, K. Zum Entscheidungsproblem der mathematischen Logik, *Verhandlungen des internationalen Mathematiker-Kongresses Zürich 1932*, vol. 2, Zurich and Leipzig: Orell Füssli, 1932, pp. 337-338.
32. Kalmar, K. Zurückführung des Endscheidungsproblems auf den Fall von Formeln mit einer einzigen, bindren, Funktionsvariablen, *Compositio Mathematica* 4, 1936, pp. 137-144.
33. Kalmar, K. On the Reduction of the Decision Problem. First Paper. Ackermann Prefix, A Single Binary Predicate, *The Journal of Symbolic Logic* 4, 1939, pp. 1-9.
34. Lindenbaum, A. Sur la simplicité formelle des notions, In *Actes du congrès international de philosophie scientifique*, vol. VII, Logique, Paris: Hermann, 1936, pp. 28-38.
35. Lindenbaum, A., and A. Tarski. Über die Beschränktheit der Ausdrucksmittel deduktiver Theorien, In *Ergebnisse eines mathematischen Kolloquiums*, fasc. 7, 1934–1935, pp. 15-22. Reproduced in A. Tarski, *Collected Papers, Vol.1 1921-1934, Vol.2*

1935-1944, Vol.3. 1945-1957, Vol.4 1958-1979, Edited by S. Givant and R. McKenzie, Birkhäuser, Basel, 1986.
and translated into English in A. Tarski, *Logic, semantics, metamathematics*, 2nd ed. (J. Corcoran), Indianapolis: Hackett, 1983. [1st ed. and transl. by J. H. Woodger, Oxford, 1956, pp. 384-39.
36. Łukasiewicz, J. *O zasadzie sprzeczności u Arystotelesa*, Kraków: Akademia Umiejętności, 1910
37. McGee, V. Logical operations, *Journal of Philosophical Logic* 25, 1996, pp. 567-580.
38. Papy, G. *Mathématique moderne*, 1-6, Paris: Didier, 1963-1967.
39. Papy-Lenger, F. and G. Papy. *L'enfant et les graphes*, Paris: Didier, 1969.
40. Purdy, R., and J. Zygmunt. Adolf Lindenbaum, Metric Spaces and Decompositions, In A. Garrido and U. Wybraniec-Skardowska (eds), *The Lvov-Warsaw School. Past and Present*, Basel: Birkhäuser, 2018, pp. 505-550.
41. Shapiro, S. (ed.). *The Limits of Logic*, Dartmouth Publishing Company, Aldershot, 1996.
42. Sher, G. *The bounds of logic*, Cambridge: MIT Press, 1991.
43. Tarski, A. Remarques sur les notions fondamentales de la méthodologie des mathématiques, *Annales de la Société Polonaise de Mathématiques* 7, 1929, pp. 270-272. English translation by R .Purdy and J. Zygmunt in J.-Y. Beziau (ed.), *Universal Logic: An Anthology*, Basel: Birkhäuser, 2012, pp. 67-68].
44. Tarski, A. *O Logice Matematycznej i Metodzie Dedukcyjnej*, Atlas, Lvov-Warsaw, 1936.
45. Tarski, A. Contributions to the theory of models. I, II, III, *Indigationes Mathematicae* 16, 1954, pp. 572-581, pp. 582-588, 17, 1955, pp. 56-64.
46. Tarski, A. *Logic, semantics, metamathematics*, 2nd ed. (J. Corcoran), Indianapolis: Hackett, 1983. [1st ed. and transl. by J. H. Woodger, Oxford, 1956].
47. Tarski, A. What are logical notions? (ed. by J. Corcoran), *History and Philosophy of Logic* 7, 1986, pp. 143-154.
48. Tarski, A. *Collected Papers, Vol.1 1921-1934, Vol.2 1935-1944, Vol.3. 1945-1957, Vol.4 1958-1979*, Edited by S. Givant and R. McKenzie, Birkhäuser, Basel, 1986. Reviewed by J. Corcoran in *Mathematical Reviews* (91h:01101, 91h:01101, 91h:01103, 91h:01104). Reprinted by Birkhäuser, Basel, 2019.

49. Tarski, A., and S. Givant. *A formalization of set theory without variable*, Providence: American Mathematical Society, 1987

50. Tarski, A., J. Tarski, and J. Woleński. Some Current Problems in Metamathematics, *History and Philosophy of Logic* 16, 1995, pp. 159-168.

51. Woleński, J. *Logic and Philosophy in the Lvov-Warsaw School*, Dordrecht: Kluwer, 1989.

52. Woleński, J. Applications of squares of oppositions and their generalizations in philosophical analysis, *Logica Universalis* 1, 2008, pp. 13-29.

53. Woleński, J. Alfred Tarski (1901-1983), In A. Garrido and U. Wybraniec-Skardowska (eds.), *The Lvov-Warsaw School. Past and Present*, Basel: Birkhäuser, 2018, pp. 361-371.

54. Woleński, J. Some Philosophical Aspects of Semantic Theory of Truth, In A. Garrido and U. Wybraniec-Skardowska (eds.), *The Lvov-Warsaw School. Past and Present*, Basel: Birkhäuser, 2018, pp. 373-389.

55. Woleński, J. Jerzy Słupecki (1904-1987), In A. Garrido and U. Wybraniec-Skardowska (eds.), *The Lvov-Warsaw School. Past and Present*, Basel: Birkhäuser, 2018, pp. 567-573.

56. Woleński, J. Adolf Lindenbaum, *The Internet Encyclopedia of Philosophy*, https://www.iep.utm.edu/lindenba/, 2020.

57. Woleński, J. The Semantic Theory of Truth, *The Internet Encyclopedia of Philosophy*, https://www.iep.utm.edu/s-truth/, 2020.

58. Zygmunt, J. Tarski's first published contribution to general mathematics, In J.-Y. Beziau, (ed.), *Universal Logic: An Anthology*, Basel: Birkhäuser, 2012, pp. 59-66.

59. Zygmunt, J., and R. Purdy. Adolf Lindenbaum: Notes on His Life with Bibliography and Selected References, *Logica Universalis* 8, 2014, pp. 285–320.

Notes

---

1. I have launched in 2019 the *World Logic Day*, celebrated in 60 locations all over the world on January 14, the day of birth of Tarski and of the death of Gödel (cf. [13]), and subsequently made the proposal to UNESCO to recognize this day. It officially entered into the UNESCO calendar of international days in 2020. Before that I managed to launch in Poland the *Alfred Tarski Prize of Logic*, part of the project *A Prize of Logic in Every Country!* (cf. [11]).

2. Each of these four volumes has been reviewed by Corcoran in *Mathematical Reviews* in 1991 (see [48]). During many years they were out of stock. They have been re-issued by Birkhäuser in 2019 [48].
3. We are preparing a volume with posthumous papers (such as the one here discussed) and correspondence (to be published also by Birkhäuser).
4. Givant wrote two interesting papers in *The Mathematical intelligencer* about Tarski for a general audience (see [27] and [28]) and there is also the book by Solomon and Anita Feferman about Tarski's life and work [23].
5. As Jan Woleński pointed out [51], the first introduction to modern logic in Poland is a presentation of Schröder's logical ideas as an appendix to Łukasiewicz's book about the principle of contradiction in Aristotle [36]. Jan Łukasiewiecz (1878-1956) was, together with Stanisław Leśniewski (1886-1939), the main teacher of Tarski.
6. Tarski also used the word "notion" in the title of his 1929 paper [43] about consequence operator (in French, but this is exactly the same word, syntactically and semantically, as in English). In this paper he presents the consequence operator as a fundamental notion of the "methodology of mathematics" which for him is here synonymous with "logic". I have recently developed a theory about *notion* (cf. [10]) in harmony with Tarski's use of this word in his 1929 paper and his 1986 paper.
7. Rohit Parikh reported that he attended a similar talk by Tarski at Bristol University (UK) at about the same period as the talk in London and Michael Dunn attended also a similar one at Rice University (Houston, USA), in January 1967. I am grateful to both of them to have informed me about that.
8. First paragraph of "L'enfant et le bon Dieu", first chapter "Rravail et découverte" of the first part of "Fatuité et renouvellement of Grothendieck's autobiography *Récoltes et Semailles* [29] (thanks to Laurent Lafforgue for the precise reference).
9. I have been quite influenced by some ideas of Lindenbaum and for this reason, I have been working at making his work better known. This has resulted in the publication of three papers about his life and work: [59], [40] and [56].
10. See [31], [32], [33]. I am grateful to Lloyd Humberstone for these references.

11. We have introduced this coloring of the square in [3]. For recent developments on the square of opposition see [15] and [16]. There is also a special issue of the journal *History and Philosophy of Logic* on the square [18].

12. Thanks to Arnon Avron who pointed out the incompleteness of a previous version of this diagram.

13. The word "variety" is used with a different meaning in Universal Algebra, cf. the famous HSP theorem [19]. But this use is rather artificial, not directly connected to the meaning of the word in natural language.

14. His main book on the subject is [51] but he published/edited lots of other books on the topic. He also edited together with the son of Tarski an interesting posthumous paper by Tarski [50].

15. This series of events was launched together with my colleague Ricardo Silvestre.

16. He was also keynote speaker at the *1st World Congress on Analogy* in Puebla, Mexico, November 4-6, 2015; an event I co-organized with Juan Manuel Campos Benítez and Katarzyna Gan-Krzywoszyńska. I remember a long discussion I had with him on the bus going back from Puebla to Mexico International Airport.

17. This book was launched at the 6th UNILOG in Vichy, France in June 2018, with the participation of Woleński.

Idea of Artificial Intelligence

Kazimierz Trzęsicki

Warsaw University of Technology,
Plac Politechniki 1,
00-661 Warsaw, Poland
e-mail: kasimir4701@gmail.com

> Nothing is more important than to see the sources of invention which are, in my opinion more interesting than the inventions themselves.
>
> G. W. Leibniz [135].

To Professor Jan Woleński for his 80th birthday in a gift.

1. Introduction

The following text was written by a man and not by a machine. Some pioneers of artificial intelligence predicted that in the 21st century machines would be "thinking." February 2019 OpenAI reported on the creation of the GPT-2 algorithm, which can write competent, reasonable essays[1]. It would be disappointing for these predictions that machines do not "publish" (all) yet. The author of this text, however, not only wrote it on a computer, he also used automated support, such as checking compliance with a dictionary, thus avoiding lexical errors. The bibliography was automatically compiled according to a given pattern from data obtained from the bibliographical database. He also used the Internet to decide what to look for, what to use and how to systematize the knowledge he acquired and to draw the conclusions that this and other knowledge had provided.

Artificial intelligence, AI, is a challenge, and as John McCarthy (1927 – 2011) believed in the 1960s, a breakthrough can occur in five to 500 years, but this challenge can never be abandoned.

The term 'artificial intelligence' (AI) was coined by John McCarthy in 1955[2] in connection with a research project. In his proposal we read [93]:

> The study is to proceed on the basis of the conjecture that every aspect of learning or any other feature of intelligence can in principle be so precisely described that a machine can be made to simulate it. An attempt will be made to find how to make machines use language, form abstractions and concepts, solve kinds of problems now reserved for humans, and improve themselves.

The Dartmouth Conference 'Summer Research Project on Artificial Intelligence' in 1956 was the first artificial intelligence conference. And there was a shift away from the physical model, the cybernetic machine thinking model, to the non-physical model, a logical, symbolic formalized system.

The term 'artificial intelligence' is one of those that can be considered a suitcase word, and therefore the initiator of this term, and the co-creator of artificial intelligence, Marvin Minsky, understands the words in which are "packed" a variety of meanings [98]. By AI, we mean both the device, the machine, and the theory of how this device works.

The context of using the term 'AI' should approximate the meaning in which it is used in a given place. The aim of AI as a field of science is to acquire knowledge that will enable the creation of AI, the assessment of the quality of operation and theoretical and practical limitations. First of all, AI is ultimately nothing more than a desire to replicate human cognitive skills in machines. The term 'artificial intelligence' could be replaced by 'cognitive technology,' which would be in substance closer to what is the subject of this discipline. AI is a research field focused on the development of systems capable of performing tasks that require human intelligence. AI as the target is a machine – it was in Alan Turing's mind, proposing a test – whose behavior is not distinguishable from human behavior [128].

The idea of what we call artificial intelligence today is – as McCorduck [94] claims, for example – rooted in the human need to do something on your own. As God created man in his likeness, so man in his likeness creates artificial intelligence. AI creators would be in this long tradition, covering everything from the time of the appearance of the Decalogue, whose first commandment prohibiting the creation of idols – you will not have other gods before me – to homunculus [14], Paracelsus (1493/4 – 1541), Golem created by Yehudah Loew ben Bezalel "Maharal" (1512/1526 – 1609) born in Poznań, Rabbi of Prague [95] and Frankenstein [27] invented by Mary Shelley (1818). However, this only points to the possible motives of those who dreamed of creating or created artificial intelligence in one form or another. These are imponderable. They are present in all human activity, and in particular in creative and scientific activity.

This consideration will be devoted to the idea of artificial intelligence and the formation of what provided a cognitive basis for scientific research or, possibly, of what is genetic to this research. So we're going to think about the intellectual rationale and the cognitive rationale of AI research. We will skip – if this does not involve the cognitive aspect in which we consider AI – the various implementations starting with the mythical products of Hefajstos, the walking lion Leonardo da Vinci [9] and others.

2. Rajmundus Lullus

The idea of artificial intelligence can already be seen at the beginning of philosophy in ancient Greece [25, pp. XV-XVII]. The inquiries of Greek philosophers, in particular the formation of the idea of formal rules of reasoning, interested one of the contemporary artists of AI, Marvin Minsky (1927 – 2008) [97, p. 106]. When the Greeks came up with logic and geometry, they were fascinated by the idea that any reasoning could be reduced to a certain kind of accounting. The greatest achievements of this ancient period include Aristotle's concept of formal logic and its syllogistics.

At the beginning of the road to artificial intelligence, however, there were dreamers. Ramon Lull (c. 1232/33 – c. 1315/16), a Catalan from Mallorca, which was then – and these were the Reconquista times, which only ended in 1492 – inhabited by large groups of Jews and Muslims. So he lived *ex orientte lux*. He is one of the most prominent writers, philosophers and scientists [10], [106].

The University of Barcelona has set up a research center on Ramona Llulla's achievements[3]. The importance of Lullus's concept for the development of artificial intelligence [28] is being considered. Lullus's legacy is also being studied at the University of Valencia. Lullus is recognized as the most influential Catalan writer and author of the first European novel *Blanquerna* [8]. The Lullus' Tree of Sciences is used as the Spanish logo of the Consejo Superior de Investigaciones Científicas (High Council for Scientific Research)[4]. The new edition of all Lullus works prepared by the Raimundus-Lullus-Institut Freiburg im Breisgau) will cover 55 volumes [122]. Recent studies show Lullus's achievements in election theory, including that he was the author – formulated a few centuries later – of the Bordy method and the Condorecta criterion. The terms 'Llull winner' and 'Llull loser' [121, chapter 3] appeared due to his works.

He is referred to as Doctor Illuminatus – a nickname he gained after meeting Duns Scotsman in 1297 – but he is not among the doctors of the Catholic Church. In 1847 he was beatified by Pope Pius IX, although in 1376 his rational mysticism was condemned by Pope Gregory XI and again by Pope Paul IV. 100 of his theses were condemned by the inquisitor Nicholas Eymerich (approx. 1316 – 1399) – yet Lullus remained in good relations with the Church. Lullus's work was synthesized by his student Thomas Le Myésier (13th century – 1336) in *Electorium* [68].

The statue of Lullus in Montserrat is characterized by the order of God – modeled on the figure of Logica Nova (1512) – by eight-step stairs: stone, flame, plant, animal, man, sky, angel, God. They symbolize the hierarchy of sciences (states of consciousness) that Lullus proclaimed. Lullus inspired many and more artificial intelligence researchers [107].

In 1265, at 33, Lullus was apprehended and became a Franciscan storyteller. He proclaimed that three religions recognizing the *Old Testament*: Judaism, Christianity and Islam should be united to stop hordes of oppression from Asia. He got involved in missionary work. He wanted to act with logic and reason. In approximately 1274 he experienced enlightenment at Mount Puig de Randa (Majorca) and got the idea of a method that he later described in the 1305 edition *Ars magna generalis ultima* [88], [91]. It was accompanied by the abbreviated version *Ars brevis* [87]. The art he designed was based on loans from Arabs – which he didn't hide – it was supposed to be a tool for converting unbelievers. Lullus spent years studying the doctrines of Jews and Arabs.

Lullus wanted to show that the Christian doctrine can be obtained mechanically with a fixed resource of ideas. One of Lullus's numerous tools for his method was the volvelle, as he called a device he had constructed.

If the logical machine is understood as the logic data processing system, Aristotle, creating the concept of formal logic, gave rise to a symbolic logical machine, and Lullus' volvelle can be seen as a physical logical machine, and this is usually referred to as a 'thinking machine.'

The name "volvelle" comes from the Latin verb "volvere", which means as much as "rotate". Inspiration can be seen in the Arabic astrological device *zairja* [85]. Lullus most likely experience of *zairja* would have been during the missionary expeditions [86], [129]. *Zairja* was used by Arab medieval astrologers.

The term '*zairjah*' derives from the Persian words '*za'icha*' (horoscope, astronomical table) and '*da'ira*' (circle) [85, p. 216].

A volvelle was made of paper or parchment. There was a volvelle with which to resolve religious disputes. A combination of nine letters was produced, representing nine attributes of God (which all monotheists recognize) written on a moving wheel. Depending on the subject, there were two or more such wheels. Another volvelle, called the "Night Sphere" by Lullus, was used to calculate the time over the night by the position of the stars. It was possible to determine the hours in which, according to the movement of the heavenly bodies, medication is most effective. The moving parts of the volvelle were placed on the blue bodies on the timer or on God's attributes and arguments for His existence, but it depended on the subject. Lullus wanted to – as if we would say today – mechanize the reasoning process. He claimed that his art lead to more certain conclusions than logic itself, and that it is therefore possible to learn more in a month than through logic in a year.

Werner Künzel was so fascinated by Lullus' 'machine' that he writes [67]:

> Since 1987, I have programmed this first beautiful algorithm of the history of philosophy into the computer languages COBOL, Assembler and C.

The Lullus method assumed that the number of fundamental truths is limited, and all the truths of a given field are derived from them in general by combinations of relevant terms. The machine was supposed to put together combinations and to indicate which ones are real.

A volvelle [112] is also a functionally related astrolabe. An astrolabe is a device that has been used to observe and calculate the positions of heavenly bodies. It can be seen as a kind of analog computer for astronomical calculations.

Volvelle, or rather those who used them, were suspected of black magic. Perhaps this approach was based on the mystical inspiration of the creator Lullus, and the fact that the device was used to predict the future. Numbers and measurements were attributed to spiritual and supra-natural potentials.

In Lullus' time, especially in Spain, the Jewish community developed a Kabbalah, and its origins take place in Cataloni in the 12th century [45], [46]. According to the Jewish tradition, Hebrew is the language that God used to create the world. The *Sefer Yetsirah* (*Book of Creation*), one of the earliest Jewish mystical texts (it was written between the 2nd and 7th century), describes the process of creation as being accomplished with 22 letters of the Hebrew language and cardinal numbers. The *Sefer Yetsirah* explained how one could imagine and possibly repeat the creation by manipulating the letters of the Hebrew alphabet. Thus, was created the Golem (*Psalm* 139:16). It was believed that by giving the name to the Golem one could revive him and control his conduct, and by wiping out that name one could destroy him.

Kabbalah interprets the Torah using anagrams and other linguistic combinations. Lullus can be seen as someone who inspires these techniques in the search for a new way of evangelization. He wrote about Kabbalah[5] that its object is creation, or language. For this reason, it is clear that its wisdom governs the other teachings. They have their roots in it. For this reason, these teachings are subordinated to this wisdom, and the principles of science and their rules are subordinated to the rules of Kabbalah. The scientific argument alone without the Kabbalah is insufficient.

Lullus provided the basis of the medieval Christian Kabbalah in its various varieties. In each case, the objective was one: by applying the rules of Kabbalah to prove that Jesus was the Messiah. Because God created the world using the Hebrew language, the contemplation of this language was the contemplation of both God and His creation. Lullus used the Latin alphabet, but the idea of the combination was the same.

Computer scientists have identified Lullus as someone who provided the (pre)origins of computer science [16], [10, p. 290], [65, p. 56]. Lullus is the one from whom you can start the story of ideas of thinking machines, which is the story of artificial intelligence.

Lullus's idea was revolutionary for two reasons, namely that the volvelle could be seen as an 'artificial memory,' which freed the user from remembering a large amount of detailed information, and its resources could be exchanged and then it could produce new knowledge. The content of this knowledge was dependent on the content of 'memory.' So in a sense, it was the idea of a universal machine.

Lullus is an important figure in the history of AI, primarily for the reason that he has interacted with many prominent researchers who have relaunched his idea in successive eras [10, pp. xii-xiv]. The idea of *ars raymundi* has revived the European public's inquiries for several centuries.

Let's list the most prominent Lullists in chronological order according to the date of their birth who contributed to the development of AI. So, we'll skip characters like Martin Luther (upon whom Lullus also acted on).

3. Lullists

Lullus gave us the beginning of a concept that has survived at least until the times of Gottfried Leibniz [81], [119]. Among many ideas, let's point out those whose ideas had the most impact on building a thinking machine. Not everything is known. In the 16th century, the biggest Lullist was Franciscan Bernard de Lavinheta. However, we do not know much about him. It is known that his release of Lullus' work was most common in Europe at the time [89, vol. I, p. 80].

3.1. Giovanni de la Fontana

Giovanni de la Fontana (c. 1390 – 1455/56) [38] was an outstanding – as we would say today – designer. He learned the art of engineering from Greek and Arabic texts. In the encrypted *Bellicorum instrumentorum liber, cum figuris et fictithousand litoris conscriptus* [50] he illustrated and described various instruments of war. In the *Secretum de thesauro experi mentorum ymaginationis hominum* [32] he made available to readers about 1430 – also written in an encrypted manner – in which he studied different types of memory and explained the function of artificial memory. He proposed some devices for remembering and 'machinery' with fixed structure and mobile parts and variables, allowing a combination of characters – including a direct link to the Lullus design.

3.2. Nicholas of Cusa

Nicholas of Cusa (1401 – 1464) in the *De coniecturis* [99] develops its method *ars generalis coniecturandi*. He describes how to make assumptions, illustrating this with circular diagrams and symbols very similar to Lullus'. Venice, in which he lived, entered into contact with Byzantine and Arab countries. The question that Lullus had asked two centuries earlier became natural about the universal language for building an agreement between East and West.

3.3. Giordano Bruno

Giordano Bruno (1548 – 1600) uses Lullus' idea to create artificial memory, and he uses this technique to make rhetorical discourse. Kircher comments later in 1669 [59, p. 4] that Giordano Bruno also developed Lullus' volvelle technique so that an unlimited number of sentences can be generated [12]. In his system, alphabetic combinations do not lead to images, but rather combinations of images lead to syllables. This system not only facilitates memory, but also enables the generation of almost unlimited words [26].

3.4. Thomas Hobbes

Thomas Hobbes (1588 – 1679) is not referred to as Lullist in the sense of referring to Lullus. The Hobbs' doctrine is important primarily because of the concept of thinking as a calculation and influence on Leibniz. I also know nothing about the contacts between the outstanding Lullist Kircher and Hobbs. Hobbes was 14 years older than Kircher. Hobbes published the *Leviathan* in 1651 that we are interested in and Kircher published the *Ars Magna Sciendi* in 1669, 18 years later [59].

Hobbes uses the term '*ratiocinari*' to mean both reasoning and accounting, as one thing. It was understood as calculation consisting of addition and subtraction, simply an arithmetic operation. He cited various reasons for this approach, referring to the meaning of the relevant words in Greek and Latin [42, chapter IV]. He added that 'syllogism' actually means adding, summing. The word count corresponds to the grammar, the syntactics of natural language, understood as an operation on words.

Hobbes is the first who directly formulated the concept of syntactic operation as calculation. Syntactic procedures are arithmetic. Hobbes recognizes the functional nature of syntactics as a kind of technical procedure. Words are used as numbers, i.e. as agreed artificial marks. His saying is famous [42, chapter IV]: "Words are wise men's counters"[6]. The symbolic character of words is, according to Hobbs, the essence of their nature from the very beginning of creation. Adam invented the words *ex arbitrtrio*. Although, as Hobbes writes [42, chapter IV]:

> The first author of Speech was GOD himself, that instructed Adam how to name such creatures as he presented to his sight.

Hobbes had a negative score on the Kabbalah. At the end of Chapter XL of the *Leviathan*, he wrote that the Kabbalah took over the Greek demon and through the Kabbalah the Jewish religion became more corrupted (their Religion became fly corrupted).

On reasoning as calculation Hobbes writes [42, chapter V]:

> When a man *reasons*, he does nothing else but conceive a sum total from addition of parcels – These operations are not incident to Numbers onely, but to all manner of things that can be added together, and taken one out of another. […] The Logicians teach the same in Consequences Of Words; adding together Two Names, to make an Affirmation; and Two Affirmations, to make a syllogisme; and Many syllogismes to make a Demonstration; and from the Summe, or Conclusion of a syllogisme, they substract one Proposition, to finde the other.

He also writes further:

> Out of all which we may define, (that is to say determine,) what that is, which is meant by this word Reason, when

wee reckon it amongst the Faculties of the mind. For Reason, in this sense, is nothing but Reckoning (that is, Adding and Subtracting) of the Consequences of generall names agreed upon, for the Marking and Signifying of our thoughts; I say Marking them, when we reckon by our selves; and Signifying, when we demonstrate, or approve our reckonings to other men.

The first task of language is a mental discourse, and therefore it is a cognitive function. The second task is to transfer knowledge to others. The third is to communicate our will to others, and the fourth is an entertainment and artistic function [42, chapter IV].

Hobbs' views on language and reasoning were significantly influenced by mechanics, the new subdiscipline of physics that Galileo Galilee provided the beginning of [132]. Galileo says: *"universum horologium est."*

For Hobbs the computational use of natural words is the first need to obtain a reasonable, i.e. a real insight, and secondly, if the calculation is done right, get complete reliability and complete confidence.

3.5. Athanasius Kircher

Athanasius Kircher (1602 – 1680) is the famous Jesuit scholar, the new Aristotle, the last who knew everything [31], the master of one hundred works [109], [110], the last man of the Renaissance [39] – he has a multitude of contributions to mnemotechnology, to the development of mechanization of calculating of "thoughts," to the design of slots and to the search for a universal language that would ultimately free humanity from the curse of the tower of Babel [82].

Kircher's scientific achievements impress with both diversity and size[7]. As a curiosity, he was the first scientist to be able to ensure his preservation from the sale of books [52, p. 96].

Findlen writes [31, p. 329]:

> During his own lifetime his books could be found in libraries throughout the world. He had a global reputation that was virtually unsurpassed by any early modern author.

In the *Encyclopedia Britannica* we read:

> [...] settled in 1634 in Rome. There he remained for most of his life, functioning as a kind of one-man intellectual clearinghouse for cultural and scientific information gleaned not only from European sources but also from the far-flung network of Jesuit missionaries.

The interest in the person and achievements of Anathasis Kircher dates back to the 1980s. For three centuries he was forgotten. Knittel (1644 – 1702) wrote the following book about Kircher in 1682: *Via Regia ad omnes scientias et artes. Hoc est: Ars universalis, scientiarum omnium artiumque arcana facili us penetrandi* [4]. It was the last thesis that openly defended Kircher's approach to knowledge, which was the subject of sharp criticism at the time. Knittel as his authority points to Pitagoras (c. 570 – c. 495 B.C.), Aristotle (384 – 322 B.C.), Raimundus Lullus, Sebastián Izquierdo (1601 – 1681), and Kircher. The *Via Regia* was very popular and had numerous editions [44]. At this time, Newton, who, like Leibniz, was fascinated by many of the questions that triggered Kircher's concept, came to completely different conclusions.

Donald Knuth in the *Art of Computer Programming* [65, pp. 60-61] points out three 17th century authors, as those who made discoveries used by computer science. They are: Tacquet, van Schooten, and Izquierdo mentioned above. Sebastián Izquierdo is the author of the work *Pharus scientiarum ubi quidquid ad cognition humanam humanitús acquisibilem pertinet, ubertim iuxtà, atque succinctè pertractaur* [49].

Today's science historians see Kircher's scientific achievements as helpful in understanding the transition from ancient to modern ways of thinking about the world [61]. Major research projects are being carried out [4], [37], [51], [123].

The Museum of Jurassic Technology[8] has a permanent exhibition dedicated to Kircher and his legacy: 'Athanasius Kircher: The World Is Bound With Secret Knots'. From 07.03 to 10.04.2008 in Collegio Romano, where Kircherianum was there, the artist Cybéle Varela organized an exhibition 'Ad Sider per Athanasius Kircher' ('To the Stars by Athanasius Kircher').

His correspondence must be taken into account when trying to determine the inspiration and influence of Kircher's work. Among the

686 people who wrote to him are, among others, Leibniz, Torricelli, and Gasendi [4]. The archive in Gdańsk contains his letters to Hevelius, and the archive of the Mazovian letter to Kochański. There are 2741 letters [51], [123]. In the context of these considerations, any correspondence with Hobbes would be interesting. I have not found any data about that correspondence. Descartes is not among the respondents (1596 – 1650).

Kircher takes Lullus' ideas first of all in the *Ars Magna* [59]. The work consists of XII books. There are books whose titles directly point to the issues of interest: III. Methodus Lulliana; IV. Ars Combinatoria.

Kircher not only discusses the Lullus concept, but also presents a new and universal Lullus method of combination concept. It seems to have the belief that Lullus' method of combination is secret and mystical, that is this is esoteric.

Kircher used the same wheels as Lullus, but differed in the choice of symbols to be combined. This notation makes a difference. He tried to produce possible combinations of all finite alphabets (not only graphic, but also mathematical). Kircher was known for his coding and decoding skills. He tried to read the hieroglyphs, he also learnt Coptic and he is the author of the first grammar of this language *Prodromus coputs sive aegyptiacus* [54], and in *Lingua aegyptica restituta* [56] he showed that Coptic is the last phase of development of the ancient Egyptian language. A more mathematical approach distinguishes his project from the Lullus project. The universal language, *lingua universalis*, not only allows you to understand everything, but also is a tool for close investigation.

The idea of binding digits to words is realized in gematry, which is a component of the Kabbalah [108]. The name derives from 'geometry.' Gematry originates in the Assyrian-Babylonian alphanumeric coding system. Others had similar ideas, including Greeks and Arabs.

Kircher not only addressed the theoretical issues of encryption and decryption, but also designed a coding and decoding machine. These and other machines, collected by Kircher, were in Kircherianum[9] [30], [31]. This is one of the first public museums in which, in addition to the artifacts obtained, he presented the many fruits of his invention, including models of robots, equipping them with speaking tubes so that the vending machine greeted visitors [40], [82], [83], [134]. In the 14th and 15th centuries, there were no shortage of designers of various kinds of machines and automata; as shown by someone like Leonardo da Vinci (1452 – 1519).

In 1649 Kircher invented the first of the brands, or cistae – these were wooden boxes that had written numbers, words, and sounds (*Arca musurgica*) [63], in general everything that can be automatically processed by a machine that combines things according to the logic defined and programmed by the inventor [64, p. 60], [96]. These bodies, as they were also called for because of their similarity to musical bodies, formed a complementary system of dissemination of encryption systems (polygraphic and steganographic) [31, p. 287].

In the museum of science history Museo Galileo[10] there is *Organum Mathematicum* [62], which Kircher designed for Prince Karl Joseph from Austria. It contained all the mathematical knowledge necessary for the prince. Simple arithmetic, geometric and astronomical calculations were made by manipulating wooden rods. It was possible to write messages with a digital code, design reinforcements, calculate the Easter date, and compose music. Although Kircher declared that obtaining mathematical knowledge would not be burdensome, many operations required mathematical fitness and memorization of long Latin poems [114][11]. *Abacus Harmonicus* (*Abacum Arithmetico-Harmonicum*), the tabularist method of creating music was described in the *Musurgia Universalis* [57], see also: [41], [119]. *Arca Musarithmica* used the aleatorical method to compose music, which is described as capable of producing millions of church anthems by a combination of selected musical phrases. Kircher's "musical" ideas are highlighted by Donald Knuth in his fourth volume the *Art of Computer Programming. Generating All Rrees. History of Combinatorial Generation* [65, pp. 52, 53, 59, 74].

Kircher in the *Polygraphia nova et universalis, ex combin atoria arte detecta* (1663) [58] designed not only polygraphy, an international language available to all, but also steganography, a secret language for encrypting messages. In creating polygraphy, Kircher used – as he himself writes – Lullus' *ars combinatoria*.

In the introduction to the *Polygraphia nova et universalis, ex combin atoria arte detecta* addressed to Emperor Ferdinand III Kircher wrote about polygraphy that all languages are reduced to one (*linguarum omnium ad unam reductio*). Anyone who uses polygraphy, even if he did not know anything other than his own speech, would be able to communicate with anyone else, regardless of their nationality. This polygraphy would be basically pasiography, i.e. a written language design or an international alphabet that would not have to be spoken.

These actions are motivated by the desire to restore humanity to the language before the mixing of languages, which is a consequence of the erection of the tower of Babel. These are ideas for realizing the human longing for the perfect language spoken by Adam and Eve in Paradise [26, pp. 196-200]. The longing to understand everyone, no matter what language he or she speaks, is also cited in the New Testament, when on the day of sending the Holy Spirit, everyone, no matter what country he or she was from or what language he or she was speaking, understood what the apostles preached, although they spoke in their own language.

Kircher's distinction between two dictionaries could be associated with modern methods of automatic translation: everything is translated into one distinguished language, and from this language only into each other. Dictionary A was used for encoding and dictionary B was used for decoding the message. For example,[12] [58, pp. 9-14]:

XXVII.36N   XXX.21N   II.5N   XXIII.8D   XXVIII.10 XXX.20

was decoded to Latin as:

Petrus noster amicus, venit ad nos.

According to Knittel, Kircher created *clavis universalis*, a universal key, opening access to the secrets of the universe [31, p. 5].

3.6. Universal Language

The 17th century is fertile in the concepts of artificial languages. A universal language was sought, understood as a language in which all courts and concepts could be expressed and, moreover, capable of accounting processing. It would be the language of invention in the sense of Hobbes.

John Wilkins (1614 – 1672), one of the geniuses of that time, had the task of creating a universal language. He knew Kircher's work [136, p. 452]. In the *Essay towards a Real Character and a Philosophical Language* (1668) [136], where he presented his concept of language, there is no mention of Hobbs – and he was, like Wilkins, an English philosopher. There is no mention of Leibniz, but his *Dissertatio de Arte Combinatoria* (1666) [69] was published two years earlier than Wilkins the *Essay towards a Real Character and a Philosophical Language* (1668). It turns out that Wilkins' precursor was Dalgarno, the author of *Ars Signorum* [20], cited by Leibniz.

Wilkins was mindful of the universal language, which would primarily facilitate an international communication of scholars. It was supposed to replace Latin, though it had a thousand-year history in the teaching of the Christian world. Latin, he declared, was difficult to learn. Unlike other projects of that time, the new universal language was supposed to be only a secondary language. *Lingua franca* could also be used for diplomacy, travel, trade and other situations [137].

The *lingua franca* scheme based on mathematical coding was published in 1630 by an English mathematician John Pell [92, p. 55]. The idea of simplifying Latin was also close to Giuseppe Peano (1858 – 1932) [53], a famous Italian mathematician who proposed Latin without flexion in the *Latino sine flexione, Interlingua de Academia pro Interlingua* (1903) [100]. In the context of our deliberations, it is worth highlighting Peano's reference to Leibniz by placing samples of his writings as a motto to individual paragraphs of his text. In 1926 'Instituto pro Interlingua' was established to continue the work. Until 1939, the Institute published the journal 'Schola et Vita' [7, p. 154].

4. Gottfried Wilhelm von Leibniz

Gottfried Wilhelm von Leibniz (1646 – 1716) was a scholar to whom many who referred, in particular Frege, who, writing *Begriffsschrift* (1879) [33], pursued the idea of universal language, *lingua characteristica* and formal calculation, *calculus ratiocinator*.

In the Leibniz concept, all the rational elements of Lullist inquiries have been accumulated. He took over Hobbes' heritage of the arithmetic philosophy of language. He developed his ideas of artificial language and symbolic systems [28].

In the letter to Hobbes of July 1670 [78, pp. 105-106], he wrote that he had read almost all of his works and that he had used them as with few others. This letter was not delivered to Hobbes and later remained only as a sketch [115].

Leibniz as a student became familiar with the late-scholastic thought of Jesuit Francisco Suárez (1548 – 1617), who enjoyed respect at Lutheran universities. The relationship between Leibniz and another Jesuit is interesting, namely Athanasius Kircher [36]. In the 'Synopsis Dissertationis De Arte Combinatoria', the *Dissertatio de arte combinatoria* (1666) [69] refers to Lullus and his art. He learned about it mainly through Kircher's work. 16 May 1670 he wrote a letter to Kircher [36, pp. 229-231] and received a reply on 23 June 28 [36, pp. 232-233]. Leibniz in the letter refers to his *Dissertatio de arte combinatoria* [69] and expresses admiration for Kircher's newly published work *Ars Magna* (1669) [59]. The value of *ars combinatoria* sees in its function as a *logica inventoria* and in the development of *scriptura universalis*. He writes about its use in the attempts to establish a new order and the basis of the system of law at that time. However, it emphasizes its fundamental function as a general basis for scientific knowledge. It was close to Kircher, who himself pointed to the important role of *ars combinatoria* for the solidifying of such different sciences as mathematics, medicine, law study, and theology. Leibniz was also interested in Kircher's writings about Egypt and China.

Leibniz's concept of thinking as a calculation takes over from Hobbs. It remains for him to determine what the units are (parcel) that Hobbes refers to as arguments of accounting operations. The concept of Lullus' art, developed in the *Dissertatio de arte combinatorial* [69], written at the age of 19, integrated with its metaphysics and philosophy of science.

The *Dissertatio de arte combinatoria* is an extended version of the PhD dissertation that was prepared before Leibniz undertook his mathematical research. The release in 1690 resumed without Leibniz's consent. Leibniz has repeatedly expressed his regret that there is a version in circulation that he considers immature.

Examples of problems to which the *ars combinatoria* are applied are issues from the law, music, the Aristotelian concept of four types of matter (presented in the form of diagram, and thus in a manner typical of Lullus), all of which is complex, and above all – from the point of view of the subject that we are interested in, but also of what has been the test of time – are applications to reasoning.

Leibniz is considered the most prominent logician from Aristotle until George Boole who published the *Mathematical Analysis of Logic: Being an Essay Towards a Calculus of Deduction Reasoning* (1847) [11], and Augustus de Morgan who pblished the

Formal Logic: or, The Calculus of Inference, Necessary and Probable (1847) [23].

Leibniz wanted the universal language to make it possible to make the rules of calculations logical. He wrote [77, p. 664]:

> At the same time this could be a kind of universal language or writing, though infinitely different from all such languages which have been proposed, for the characters and the words themselves give directions to reason, and the errors – except those of fact – would be only mistakes in calculation. It would be very difficult to form or invent this language or characteristic but very easy to learn it without any dictionaries.

In the letter to the mathematician G. F. A. L'Hospital, we read [22, chapter 1] that the part of the "algebra" secret is included in the characteristics, i.e. in the art of proper use of symbolic expressions. A concern for the proper use of the symbol would be *filium Ariadne*, which would lead the researchers in creating this characteristic.

In the *Dissertatio de arte combinatoria* he criticized Lullus' 'alphabet' as limited and proposed an alternative, extended, and instead of letters he considered it appropriate to use numbers. For example, he proposed that '2' should represent space, 'between' should be represented by '3' and the whole by '10'. This encoding encodes 'episode' as 2.3.10. By digital encoding, all problems will be reduced to mathematical problems and solved by accounting operations. This idea anticipates the modern AI [28]. Digital coding has already been used by other Lullists of Leibniz's predecessors.

When we proclaim the researcher's contribution to scientific development, we take into account what Leibniz knew when he wrote [77, p. 664]:

> [...] Besides taking care to direct my study toward edification, I have tried to uncover and unite the truth buried and scattered under the opinions of all the different philosophical sects, and I believe I have added something of my own which takes a few steps forward.

Leibniz's contribution to the development of the AI concept is noted, first of all, in two new novelties of his inquiries, or rather – which would be more cautious given that one can find predecessors – in indicating relevance and subsequent impact, first of all, in a situation where our knowledge is not certain and we have to settle for probability and, second, not only cognitive, but also ontological location of the binary system.

AI is supposed to behave like a man who doesn't make a mistake. AI must therefore also deal with situations that human beings deal with, in particular when taking decisions and acting in conditions of incomplete or uncertain information. This aspect is noted by Leibniz (in relation to the universal language, which in the context of his speech we can understand as a "thinking machine"). Leibniz [77, p. 664] wrote:

> When we lack sufficient data to arrive at certainty in our truths, it would also serve to estimate degrees of probability and to see what is needed to provide this certainty. Such an estimate would be most important for the problems of life and for practical considerations, where our errors in estimating probabilities often amount to more than half […]

Leibniz in many texts and letters written between 1679 and 1697, i.e. for eighteen years, developed a notation and solved an algorithmic (mechanical) execution of arithmetic operations. He also drew up a draft of rules for the binary machine, using balls and holes, sticks and grooves to move them[13] [70], [72], [116], [126], [127].

Leibniz considered the idea of three-valued logic in the *Specimina Iuriss III* [113, 1931, p. 20].

The binary system as the basis of machine counting is also indicated by the prominent English inventor Thomas Fowler (1777 – 1843), who also designed a wooden 'computer,' operating according to the rules of ternary system[14] [131].

In January 1697 Leibniz, with his birthday wishes, sent the letter to his protector Prince Rudolf Augusta of Brunswick (Herzog von Braunschweig-Wolfenbüttel Rudolph August), discussing the binary system and the idea of creation with 0 as nothingness and 1 as God [120].

For Leibniz [71], nothingness and darkness correspond to zero, while the radiant spirit of God corresponds to one. For he thought that all combinations arose from unity and nothingness, which is similar to when it was said that God had done everything out of nothing and that there were only two principles: God and nothingness. He designed a medal, whose main theme was *imago creationis* and *ex nihil ducendis Sufficit Unum*. One corresponds to the Sun, which radiates to the shapeless earth, zero. He referred to Pythagoras and Plato. From the spirit it was Kabbalistic, it was embedded in gematry.

The idea of binary code is not new [84]. Leibniz himself pointed to the predecessor in the person of the thirteenth-century Arabic mathematician Abdallah Beidhawy. In approximately 1600 the binary notation was used by the English astronomer Thomas Harriot. Shirley writes about his achievements [118]:

> Though it is frequently stated that binary numeration was first formally proposed by Leibniz as an illustration of his dualistic philosophy, the mathematical papers of Thomas Hariot (1560 – 1621) show clearly that Harriot not only experimented with number systems, but also understood clearly the theory and practice of binary numeration nearly a century before Leibniz's time.

A similar opinion is given by [47]:

> He is probably the first inventor of the binary system, as several manuscripts in his legacy show. In the binary system, he uses the numerals 0 and 1 and shows examples of how to move from the decimal system to the binary system and vice versa (conversion or reduction). Using further examples, he demonstrates the basic arithmetic operations.

Ineichen had the first publication on the binary system, in 1670. Two-volume book *Mathesis biceps vetus et nova* (1670) [48] by Ioannis Caramuelis. Either way, Leibniz developed a binary system, which is how to perform both arithmetic operations – as he described it – and logical operations – as Boole did. With his conviction that everything is created from 0 and 1, he anticipated what modern computer science is doing, that all information can be written in binaries. The ontological thesis about the world as created by 1 using 0 opened up new perspectives for linking the information system to metaphysics. While praising his binary arithmetic Leibniz claimed [79]:

> *tamen ubi Arithmeticam meam Binariam excogitavi, antequam Fohianorum characterum in mentem venirent, pulcherrimam in ea latere judicavi imaginem creationis, seu originis rerum ex nihilo per potentiam summae Unitatis, seu Dei.*

> But when I invented my binary arithmetic, before I became familiar with the symbols of Foha, I recognized in them the most beautiful image of creation, that is, the origin of things from nothing thanks to the highest power of Unity, that is, God.

This idea of Leibniz was so fascinating that it was passed on to Father Grimaldi, a mathematician at the of court of the Emperor of China, in the hope that it would lead to the conversion of the Emperor and, with him, to the Christianization of the whole of China [71].

After 1703, i.e. after the publication of *Explication de l'arithmétique binaire, qui se sert des seuls caractères 0 et 1, avec des remarques sur son utilité, et sur ce quélle donne le sens des anciennes figures Chinoises de Fohy* [72], there is an increase of interest in systems that are not decimal. The use of binary in computers was ultimately determined only by the *Burk-Goldstine – Von Neuman Report* of 1947, in which we read [13, p. 105]:

> An additional point that deserves emphasis is this: An important part of the machine is not arithmetical, but logical in nature. Now logics, being a yes-no system, is fundamentally binary. Therefore, a binary arrangement of the arithmetical organs contributes very significantly towards a more homogeneous machine, which can be better integrated and is more efficient.

Giuseppe Peano (1858 – 1932) designed an abstract shorthand machine based on the binary encoding of all Italian syllables between 1887 and 1901. Together with phonemes using 16 bits (so it had 65,536 combinations), 25 letters of the (Italian) alphabet and 10 digits were encoded. Peano's code was not noticed and was forgotten. The American Standard Code for Information Interchange (ASCII) and its various extensions are used in today's coding computers. Since 2007 coding on the Internet is done using UTF-8, which is backwards compatible with ASCII.

The idea that everything is created from 0 and 1 is the reason why the creator of the algorithmic theory of information Chaitin – as he writes not quite seriously – proposes to name the basic unit of information not 'bit' but 'leibniz' [15], [125]:

> [...] all of information theory derives from Leibniz, for he was the first to emphasize the creative combinatorial potential of the 0 and 1 bit, and how everything can be built up from this one elemental choice, from these two elemental possibilities. So, perhaps not entirely seriously, I should propose changing the name of the unit of information from the bit to the leibniz!

The 'leibniz' unit could be the unit (parcel) that Hobbes wrote about. Leibniz was convinced that the world was designed according to the principles of mathematics. This thought is abbreviated [78, p. 191]:

> *Cum Deus calculat et cogitationem exercet, fit mundus*
>
> When God thinks about things and accounts, the world appears.

Mathematics is the tool of the Constructor of the world, and numbers are the material from which the world is made. This idea is based on the Old Testament *Book of Wisdom* (canonical for Catholics and Orthodox Christians, Ethiopian and Syrian Christians – it was created in the Hellenistic world), in which we read (11:20):

> But you have arranged all things by measure and number and weight!

The idea of world mathematics lies at the heart of modern natural science, the origins of which are usually related to the speech of Galileo, who claimed that the book of nature is written in the language of mathematics.

If thinking is a calculation, and the world is made of numbers, then we will come to any truth that we can come to, by the way of accounting. Thus [75, vol. 7, p. 200][15]:

> *Quo facto, quando orientur controversiae, non magis disputatione opus erit inter duos philosophos, quam inter duos Computistas. Sufficiet enim calamos in manus sumere sedereque ad abacos, et sibi mutuo (accito si placet amico) dicere: c a l c u l e m u s.*

> If the dispute had arisen, the dispute between the two philosophers would not have required much effort than between the two accountants. For it would be sufficient for them to take pencils into their hands, to sit by their slats, and one to the other (with a friend as a witness if they wished) to say: Let's count.

Calculating is an activity in which a machine can replace a human. In 1685, in discussing the value for astronomers of a machine invented in 1673 more efficient than pascalina and performing all basic arithmetic activities, he wrote [22, chapter I: Leibniz's Dream], [76, p. 181] that:

> For it is unworthy of excellent men to lose hours like slaves in the labor of calculation which could safely be relegated to anyone else if the machine were used.

This pragmatic argument with the above metaphysical arguments can inspire computer science and the development of its tools towards artificial intelligence. All truths have a numerical representation, and thinking is represented by numerical operations, and all this can be done by the machine.

Frege critically continues the Leibnizian program, as he writes in the introduction to the published *Begriffsschrift* [33], [34, p. XI]:

> Auch Leibniz hat die Vortheile einer angemessenem Bezeichnungsweise erkannt, vielleicht überschätzt. Sein Gedanke einer allgemeinen Charakteristik, eines calculus philosophicus oder ratiocinator war zu riesenhaft, als dass Versuch ihn zu verwirklichen über die blossen Vorbereitungen hätte hinausgelangen können. Die

Begeisterung, welche seinen Urheber bei der Erwägung ergrift, welch unermessliche Vermehrung der geistigen Kraft der Menschheit aus einer die Sachen selbst treffenden Bezeichnungsweise entspringen würde, liess ihn die Schwierigkeiten zu gering schätzen, die einem solchen Unternehmen entgegenstehen.

Wenn aber auch dies hohe Ziel mit Einem Anlaufe nicht erreicht werden kann, so braucht man doch an einer langsamen, schrittweisen Annäherung nicht zu verzweifeln. Wenn eine Aufgabe in ihrer vollen Allgemeinheit unlösbar scheint, so beschränke man sie verläufig; dann wird vielleicht durch allmähliche Erweiterung ihre Bewältigung gelingen. Man kann in den arithmetischen, geometrischen, chemischen Zeichen Verwirklichungen des Leibnizischen Gedankens für einzelnen Gebiete sehen. Die hier vorgeschlagene Begriffsschrift fügt diesen ein neues hinzu und zwar das in der Mitte gelegene, welches allen anderen benachbart ist. Von hier aus lässt sich daher mit der grösten Aussicht auf Erfolg eine Ausfüllung der Lücken der bestehenden Formelsprache, eine Verbindung ihrer einzigen und eine Ausdehnung auf Gebiete ins Werk setzen, die bisher einer solchen ermangelten.

Leibniz also recognized the advantages of a suitable method of labeling, perhaps overestimated by him. His idea of universal characterization, *calculus philosophicus* or *ratiocinator*, was too titanic, so that the attempt to make it a reality could only be achieved by preparation. The enthusiasm which took over his initiator in considering how it unimaginably multiplied the spiritual power of mankind, which would in fact flow from the proper way of marking, made it estimate the difficulties too weakly that such an undertaking would encounter. When they did not reach the target at one time, they should not be doubted as they approached slowly in steps.

When a task in its entirety seems insoluble, it is temporarily restricted; then, perhaps, through a gradual enlargement, it will be resolved. Arithmetic, geometric, and chemical signs can be seen as the realization of Leibniz's idea for these particular fields. Here, the proposed conceptual letter supplements them with new ones and, although it is in the middle, what is close to everyone else. Hence, it seems to have the biggest view of the success of filling this gap in the existing formula language, by developing a combination of the individual and extending to the areas that lacked it.

There's no idea of using a language designed by Frege in learning. *Lingua universalis* brings us closer to programming languages. John McCarthy, one of the initiators of modern AI research, created the LISP[16] programming language. Today LISP is a family of such languages.

Leibniz was not only interested in the Kabbalah, but the concepts of Kabbalah, especially those of Lurian, had an impact on his views and actions mainly thanks to Franciscus Mercurius van Helmont (1614 – c. 1698/1699), who was a frequent visitor in Hanover and with whom Leibniz spent much time. He had already learnt the Kabbalah as a student. In the 17th century, in the times of the Enlightenment, Platonism, Kabbalism, and Gnosticism were popular, especially in Protestant Germany. In the case of ecumenical Christians like van Helmont, the Kabbalah had a significant impact on their optimistic non-dogmatic philosophy [18]. Leibniz, at the end of his life, accepted the radical Kabbalistic idea of *tikkun*, and the belief was that all things would ultimately be perfected by recurring transformations.

He believed in progress. He was involved in efforts to improve human health through ecumenical action, the promotion of tolerance, and the development of education and science. Leibniz's attitude to knowledge was expressed by the *theoria cum praxis* formula, which is the motto of the Kurfürstlich Brandenburgischen Sozietät der Wissenschaften (now: Berlin-Brandenburgischen Akademie der Wissenschaften). Leibniz-Sozietät der Wissenschaften[17] uses the motto: *theoria cum praxis et bonum commune*. He claimed that if we consider disciplines in and for ourselves, they are all theoretical; if we consider them from the point of view of application, they are all practical.

Socially useful ideas were also meant to improve life. He was very interested in various kinds of inventions, for example. He corresponded with Papin, who was building a steam machine – which Frege comments on later [35]. Leibniz [76] is known as the designer of the calculating machine. He had the idea since 1672. The first structures, as the documents show, took place between 1674 and 1685. The so-called older machine was made in the years 1686 – 1694. The younger machine, which behaved, was built in the years 1690 – 1720.

In Göttingen in 1879, the original of the instrument was found. One of the copies which he had constructed Leibniz had given to Peter the Great and the latter gave it to the emperor of China. Leibniz designed a high-speed car that would travel along the road like a ball bearing, designed drainage in Hartzu mines, a navigation system, utilization of wasted heat furnaces, tax reform, public health services, including epidemic-related, fire protection, steam fountains, street lighting, and state bank. He was even interested in mundane matters such as wheelbarrows or cooking pots. He designed shoes with springs so that he could walk faster. These ideas and projects were considered in the company of van Helmont.

Leibniz can be considered the last one for whom Lullus' ideas were the direct inspiration of their philosophical concepts and which proved to find a permanent place in the history of science and philosophy.

5. Forgotten Scholars

Even though it may be assumed that Kircher's project knowledge is not taking Leibnizian "thinking machines" as the Lullists understood them. Yes, he built a counting machine with new technical solutions compared to Pascalina. He designed a binary computer. Despite many other ideas, there is no device that would implement Lullus' ideas, as was the case with Kircher. Does he think that the function of the "thinking" machine will be taken over by the counting machine, for which he had a theoretical basis? And that only such a machine will be fit for the purposes that could be served by *ars combin atoria*?

Leibniz seems to have only pragmatic designs, as it was with Pascalina, which Pascal built to facilitate the work of his father, a tax collector, so Leibniz worked to improve human health. Even the famous "Calculemus!" can be interpreted as a tool for achieving social consensus, which was one of the goals that Leibniz set for himself.

Grimaldi, a Jesuit mathematician at the court of the Emperor of China, informed him with a fascinating binary system in the hope that with it he would lead to the conversion of the Emperor and, with him, to the Christianization of all of China [71]:

> Daher, weilen ich anitzo nach China schreibe an den Pater Grimaldi, Jesuiter Ordens, Präsidenten des mathematischen Tribunals daselbst, mit dem ich zu Rom

bekannt worden, und der mir auf seiner Rückreise nach China, von Goa aus, geschrieben; so habe gut gefunden, ihm diese Vorstellung der Zahlen mitzutheilen, in der Hoffnung, weilen er mir selbst erzählet, daß der Monarch dieses mächtigen Reichs ein sehr großer Liebhaber der Rechenkunst sey, und auch die europäische Weise zu rechnen, von dem Pater Verbiest, des Grimaldi Vorfahr, gelernet; es möchte vielleicht dieses Vorbild des Geheimnisses der Schöpfung dienen, ihm des christlichen Glaubens Vortrefflichkeit mehr und mehr vor Augen zu legen.

Therefore, because I am writing to China to Father Grimaldi, of the Jesuit Order, the chairman of the mathematical college of the same one with whom I met in Rome, and who wrote to me on the way back to China, from Goa; so I thought it appropriate to inform him of this presentation of figures, in hope, because he himself told that the monarch of this powerful empire is a very great enthusiast of the art of accounting, and also from father Verbiest, the predecessor of Grimaldi, who learned the European way of accounting; that perhaps this depiction of the mystery of creation could serve to give him the ever more glorious Christian faith first hand.

In the Leibniz era, Athanasius Kircher realized the most successful AI project. This theory does not in any way detract from Leibniz's scientific and philosophical achievements. It belongs to those thinkers to whom are sometimes attributed more. An example is the case of Leibniz's contribution to the development of modern logic. According to Peckhaus [105]: The development of modern logic in the UK and Germany in the second half of the 19th century can only be explained as an unconscious first, and only later a conscious reference to the Leibnizian program. Hence, the assessment of the importance of Leibniz's logic for the development of modern logic must be greatly relativized. In another previous work, Peckhaus wrote [103, p. 436]:

The development of the new logic started in 1847, completely independent of earlier anticipations, e.g., by the German rationalist Gottfried Wilhelm Leibniz (1646 – 1716) and his followers [104], [102, ch. 5].

The question is why Kircher's work has been forgotten. A similar question can also be posed in the case of Leibniz, who was already forgotten during his lifetime, reflected in that his funeral was attended only by a personal secretary. Although he was a member of the Royal Society and Königliche-Preußische Akademie der Wissenschaften, none of these institutions honored him in any way in connection with his death, and his grave remained forgotten for more than 50 years.

Athanasius Kircher had a Catholic funeral, which was solemn. His heart was deposited in a church in Santuario della Mentorella. In 1661 Kircher found the ruins of that church, which he thought was from the days of Constantine. Kircher, by his own accord, had it rebuilt. What caused Kircher's to be forgotten for three centuries? How did it happen that "a giant among seventeenth-century scholars" and "one of the last thinkers who could rightfully claim all knowledge as his domain" [19, p. 68] fell into oblivion for three centuries?

Descartes declared Kircher more a charlatan than a wise man and as someone with an aberrational imagination. The pretext for such opinions was Kircher's description of the experiment with plant heliotropism, which apparently was not understood by Descartes. Kircher pointed to the magnetic link between the Sun and plants by experimenting with a sunflower floating in the water on cork. When the flower was spinning behind the Sun, the clue indicated the time. Kircher, as the reason for the inaccuracy indicated, blocked the attracting light through glass, which protected against the inaccuracy that the wind could cause. Descartes interpreted Kircher's description as referring to earlier speculation that attributed the heliotropic properties of sunflower seeds floating in a cup of scale. Although Kircher described experiments with other heliotropic plants, Descartes stayed at his side and launched an unbridled attack on Kircher. Descartes' authority in the emerging science according to a rational paradigm was so great that Kircher's reputation was permanently damaged. Even Nicolas-Claude Fabri de Peiresc (1580 – 1637), a longstanding supporter of Kircher, became suspicious. Despite his criticism, Kircher maintained his version of the sunflower clock, occasionally modifying and demonstrating its proper functioning. In the *Magnes, sive de arte Magnetica* (1641) [1] he noticed that this kind of clock works only a month, even when it is nurtured with the greatest care – nothing is perfect in every aspect.

In the *Mundus subterraneus* (1678) [60] Kircher writes about various creatures that live underground, including dragons, in which he believed himself as the last scholar. Rationalists are less spontaneous, but Kircher was also on the right track to recognize microbes as the cause of disease, to discover the rules of volcanism and even to formulate some prototheory of evolution.

Huygens in the letter to Descartes [24, vol. III, p. 802] of January 7, 1643 makes a marginal and disrespectful mention of Kircher's magnet[18]. In response Descartes reads [24, vol. III, pp. 803-804]:

> Je sais bien que vous n'avez point affaire de ces gros livres, mais affin que vous ne me blasmiez pas d'employer trop de temps à les lire, je ne les ai pas voulu garder d'avantage. J'ai eu assez de patience pour les feuilleter, et je croy avoir vû tout ce qu'ils contienent, bien que je n'en aie gueres leu que les titres et les marges.
> Le Jesuite a quantité de forfanteries, il est plus charletan que sçavant. Il parle entre autres choses d'une matière, qu'il dit avoir eu d'un marchand Arabe, qui tourne nuit et jour vers le soleil. Si cela etait vrai la chose serait curieuse, mais il n'explique point quelle est cete matière. Le pere Mersenne m'a ecrit autrefois, il y a environ 8 ans, que c'etait de la graine d'heliotropium, ce que ie ne crois pas, si ce n'est que cete graine ait plus de force en Arabie qu'en ce païs, car ie fus assez de loisir pour en faire l'experience, mais elle ne reussit point. Pour la variation de l'aimant, i'ai toujours cru qu'elle ne procedait que des inégalitez de la terre, en sorte que l'aiguille se tourne vers le coté oú il y a le plus de la matiere qui est propre à l'attirer: et parce que cete matière peut changer de lieu dans le fonds de la mer ou dans les concavites de la terre sans que les hommes le puissent savoir, il m'a semblé que ce changement de variation qui a eté observé à Londres, et aussi en quelques autres endroits, ainsi que raporte votre Kircherus, etait seulement une question de fait, et que la philosophie n'y avait pas grand droit.

I know you have nothing to do with these books, but because you don't blame me for spending too much time reading them, so I didn't want to keep them anymore. I had enough patience to review them, and I think I've seen everything they contain, even though I've only drawn attention to their titles and indications on the margins. This Jesuit has a lot of child in him and is more a charlatan than a scholar. Among other things, he talks about an issue he claims he received from an Arab merchant who turns day and night toward the Sun. If that were the case, the matter would be interesting, but it does not explain at all what this is about.

My father Mersenne wrote to me in the past, about eight years ago, that these are heliotropic seeds, which I don't believe, except that this grain has more strength in Arabia than in this country, because I had enough time to do the experiments, but I didn't. As for the deflection of the magnet, I always thought it was only from the unevenness of the earth, so that the needle rotates in the direction where the most matter is, which is suitable to attract it; and because this matter could change its place on the seabed or in the concavities of the earth, which people cannot know, it seemed to me that this shift in deflection observed in London, and also in several other places, as Kircher reports, was only a matter of fact, and that whole philosophy had little to do with it.

Kircher knew Descartes' opinion. A. Baillet, the biographer of Descartes [24, vol. IV, p. 413] writes:

> Le Pére Kircher ne fut pas long-temps sans changer de sentiment à l'égard de M. Descartes, dont il rechercha l'amitié par la médiation du P. Mersenne; et M. Descartes, outre des compliments et des recommandations de lui, reçût encore ce qu'il avait écrit de la nature et des effets de l'aiman, et y fit quelques observations qui se sont trouvées aprés sa mort parmi ses papiers.

Father Kircher soon changed his feelings to Descartes and via father Mersenne sought friendship with him; but Descartes, in addition to compliments and advice given to him, continued to sustain what he wrote about the nature and operation of the magnet, and made some observations that were found after his death among his documents. One more negative review is included in the letter to Colvius [24, vol. IV, p. 718]:

> Il y a longtemps que j'ai parcouru Kirkerus; mais je n'y ai rien trouvé de solide. Il n'a que de forfanteries à l'italiene, quoi qu'il soit Allemand de nation.
>
> It's been a long time since I've read Kircher, but i didn't find anything solid there. There is nothing there except childish tricks of Italian, although he is German.

Perhaps not only Descartes' opinion, but also the spirit of the age contributed. Also Descartes, who was another Jesuit educator, equated Jesuit intellectualism with the Inquisition that imprisoned Galileo and sentenced Giordano Bruno [52, pp. 95-96].

Why, four centuries after Kircher's birth, was there interest in his person and creativity? Is it because of eccletism and some similarity to postmodern thinking? [39, p. 272] explains a reason:

> his effort to know everything and to share everything he knew, for asking a thousand questions about the world around him, and for getting so many others to ask questions about his answers; for stimulating, as well as confounding and inadvertently amusing, so many minds; for having been a source of so many ideas—right, wrong, half-right, half-baked, ridiculous, beautiful, and all encompassing.

6. Conclusions

With the person and achievements of Gottfried Leibniz, the time of shaping the idea of artificial intelligence is over, and the history of artificial intelligence begins. From Leibniz the way leads to Turing not only when it comes to the universal computer [21], but also when it comes to artificial intelligence. Leibniz believed in its implementation. He wrote [77, p. 664]:

> I should venture to add that if I had been less distracted, or if I were younger or had talented young men to help me, I should still hope to create a kind of universal symbolistic [spécieuse générale] in which all truths of reason would be reduced to a kind of calculus.

The development and applications of AI change our lives as Leibniz wanted, when he wrote that it would be (*characteristica universalis*) the last effort of the human spirit, because when the project is implemented, the human tool will have the ability to expand the possibilities of reason, just like a telescope that removes vision and a microscope that enabled us to see the interior of nature.

Thanks to it, 'Leibniz an Heinrich Oldenburg' [80, pp. 373-381]:

> [...] inter loquendum ipsa phrasium vi lingua mentem praecurrente praeclaras sententias effutient imprudentes, et suam ipsi scientiam mirantes, cum ineptiae sese ipsae prodent nudo vultu, et ab ignarissimo quoque deprehendentur.

> [...] while speaking, with the very power of wording, when the tongue is guided by the mind, even the fools will speak very intelligent sentences, wondering at their knowledge, without difficulty defeating their mental inability, and even the most stupid will understand these words.

We now come to make the judgment that Leibniz called for when he wrote, 'Leibniz an Heinrich Oldenburg' [80, pp. 373-381]:

> Quantam nunc fore putas felicitatem nostram si centum ab hinc annis talis lingua coepisset.

It means:

> Judge how fortunate our happiness will be if, in a hundred years from now, such a language will arise.

For his *machine arithmetica* Leibniz designed a medal with the inscription [3, pp. 307-308][19]:

SUPRA HOMINEM
— better than mankind.

However, today as artificial intelligence becomes more realistic, it raises more fears than hopes.

References

1. *A Botanical Clock from Magnes, sive de arte Magnetica*, http://www.mjt.org/exhibits/sunflower.htm [24.01.2020].
2. *A Student of Athanasius Kircher, Gaspar Schott Publishes Treatises on the Wonders of Scientific Innovation*, http://www.rarebookroom.org/pdfDescriptions/schioc.pdf [02.02.2020].
3. *Ars Magna (Ramon Llull)*, World Heritage Encyclopedia, http://self.gutenberg.org/articles/eng/Ars_Magna_(Ramon_Llull) [24.01.2020].
4. *Athanasius Kircher at Stanford*, http://kircher.stanford.edu/ [26.01.2020].
5. Berka, K., and L. Kreiser (Eds.), *Logik-Texte. Kommentierte Auswahl zur Geschichte der modernen Logik*, Berlin: Akademie-Verlag, 1971.
6. Berka, K., and L. Kreiser (Eds.), *Logik-Texte: kommentierte Auswahl zur Geschichte der modernen Logik*, 2nd ed., Berlin: Akademie Verlag, 1973.
7. Blanke, D. *International Planned Languages: Essays on Interlinguistics and Esperantology*, S. Fiedler and H. Tonkin (Eds.), New York: Mondial, 2018.
8. *Blanquerna (Ramon Llull, 1283)*, World Heritage Encyclopedia, http://self.gutenberg.org/articles/Blanquerna [25.01.2020].
9. Block, C. *The Golem: Legends of the ghetto of Prague*, New York: Rudolf Steiner Press, 1925.
10. Bonner, A. *The Art and Logic of Ramon Llull: A User's Guide*, Vol. XCV, Leiden, Boston: Brill Academic Pub, 2007.
11. Boole, G. *The Mathematical Analysis of Logic: Being an Essay towards a Calculus of Deduction Reasoning*, Cambridge, London: Macmillan, Barclay and Macmillian, 1847; http://www.gutenberg.org/ebooks/36884 [25.01.2020].

12. Bruno, G. *De lampade combinatoria lulliana: Ad infinitas propositiones et media invenienda*, Witebergae: Welack, Matthäus, 1587.
13. Burks, A. W., H. H. Goldstine, J. von Neuman. Preliminary discussion on the logical design of an electronic computing instrument, [in:] W. Aspray and A. Burks (Eds.), *Papers of John von Neumann on Computing and Computer Theory*, Vol. 12, MIT Press, 1987, pp. 97–142; https://archive.org/details/papersofjohnvonn00vonn [25.01.2020].
14. Campbell, M. B. Artificial men: alchemy, transubstantiation, and the homunculus, *Republics of Letters: A Journal for the Study of Knowledge, Politics, and the Arts*, 1(2), 2010, pp. 4–15; https://arcade.stanford.edu/sites/default/files/article_pdfs/roflv01i02_02campbell_comp3_083010_JM_0.pdf [25.01.2020].
15. Chaitin, G. J. *Leibniz, Randomness & the Halting Probability*, http://www.cs.auckland.ac.nz/CDMTCS/chaitin/turing.html [25.01.2020].
16. Copleston, F. C. *The history of Philosophy*, Vol. 4. From Descartes to Leibniz, New York, London, Toronto, Sydney, Auckland: Image Books. Doubleday, 1994.
17. Coudert, A. P. *Leibniz and the Kabbalah*, Dordrecht: Kluwer Academic Publishers, 1995.
18. Coudert, A. P. *The Impact of the Kabbalah in the Seventeenth Century: The Life and Thought of Francis Mercury van Helmont (1614–1698)*, Leiden: BRILL, 1999.
19. Cutler, A. *The seashell on the mountaintop: A story of science sainthood, and the humble genius who discovered a new history of the earth*, New York: Dutton, 2003.
20. Dalgarno, G. *Ars signorum, vulgo character universalis et lingua philosophica*, London: J. Hayes, 1661.
21. Davis, M. *The Universal Computer: The Road from Leibniz to Turing*, New York: W. W. Norton & Company, 2000.
22. Davis, M. *Engines of Logic: Mathematicians and the Origin of the Computer*, New York: W. W. Norton & Company, 2001.
23. De Morgan, A. *Formal Logic: or, The Calculus of Inference, Necessary and Probable*, London: Taylor and Walton, 1847; https://archive.org/details/formallogicorthe00demouoft/page/n6/mode/2up [25.01.2020].

24. Descartes, R. *Oeuvres de Descartes*, Vols. 1–12 + supplement: 1913; C. A. et Paul Tannery (Ed.), Paris: Léopold Cerf. 1897–1910;
https://archive.org/details/uvresdedescartes01desc/mode/2up [25.01.2020].
25. Dreyfus, H. L. *What Computers Can't Do: A Critique of Artificial Reason*, New York: Harper & Row, 1972;
https://archive.org/details/whatcomputerscan017504mbp [25.01.2020].
26. Eco, U. *W poszukiwaniu języka uniwersalnego*, Warszawa: Volumen, 2002.
27. Evert, W. *Frankenstein: Four talks delivered on WQED-FM*, Pittsburgh: Penn, 1974.
28. Fidora, A., C. Sierra, S. Barberà, M. Beuchot, E. Bonet, A. Bonner, G. Wyllie. *Ramon Llull: From the Ars Magna to Artificial Intelligence*, Barcelona: Spanish Council for Scientific Research, 2011.
29. Findlen, P. Scientific spectacle in baroque Rome: Athanasius Kircher and the Roman College Museum, *Roma Moderna e Contemporanea*, 3(3), 1995, pp. 625–665.
30. Findlen, P. *Possessing Nature. Museums, Collecting, and Scientific Culture in Early Modern Italy*, Berkley: University of California Press, 1996.
31. Findlen, P. (Ed.), *Athanasius Kircher: The last Man Who Knew Everything*, New York and London: Routledge, 2004.
32. Fontana, J. *Methoden des Erinnerns und Vergessens: Johannes Fontanas Secretum de thesauro experimentorum ymaginationis hominum*, Vol. 68, Stuttgart: Franz Steiner Verlag, 2016.
33. Frege, G. *Begriffsschrift, eine der arithmetischen nachgebildete Formelsprache des reinen Denkens*, Halle: Verlag von Louis Nebert, 1879;
http://gallica.bnf.fr/ark:/12148/bpt6k65658c [25.01.2020];
http://dec59.ruk.cuni.cz/~kolmanv/Begriffsschrift.pdf [25.01.2020].
34. Frege, G. *Begriffsschrift und andere Aufsätze*, Hildesheim: Georg Olms Verlagsuchhandlung, 1964;
archive.org/details/begriffsschriftu0000freg/mode/2up [25.01.2020].

35. Frege, G. Über den Briefwechsel Leibnizesens und Huygens mit Papin, [in:] I. Angelelli (Ed.), *Begriffschrift und andere Aufsätze*, Hildesheim: Georg Olms Verlagsuchhandlung, 1964, pp. 93–96.
36. Friedländer, P. Athanasius Kircher und Leibniz: Ein Beitrag zur Geschichte der Polyhistorie im XVII. Jahrhundert. Rendiconti, *Atti della Pontificia Accademia Romana di Archeologia*, 13, 1937, pp. 229–247.
37. *GATE, Gregorian Archives Texts Editing*, https://gate.unigre.it/mediawiki/index.php/Gregorian_Archives_Texts_Editing_(GATE) [16.01.2020].
38. *Giovanni Fontana*, https://history-computer.com/Dreamers/Fontana.html [25.01.2020].
39. Glassie, J. *A man of Misconceptions. The Life of an Eccentric in an Age of Change*, New York: Penguin Random House, 2012.
40. Gorman, M. J. Between the demonic and the miraculous: Athanasius Kircher and the baroque culture of machines, [in:] D. Stolzenberg (Ed.), *Encyclopedia of Athanasius Kircher*, Stanford: Stanford U. Libraries, 2001, pp. 59–70; docshare.tips/athanasius-kircher-and-the-baroque-culture-of-machines_57771ea9b6d87ff9378b49c1.html [25.01.2020].
41. Gouk, P. Making music, making knowledge: the harmonious universe of Athanasius Kircher, [in:] D. Stolzenberg (Ed.), *Encyclopedia of Athanasius Kircher*, Stanford: Stanford U. Libraries, 2001, pp. 71–83.
42. Hobbes, T. *Leviathan or the matter, forme, & power of a commonwealth ecclesiastical and civill*, Green Dragon in St. Paul's Churchyard: Andrew Crooke, 1651; http://www.gutenberg.org/ebooks/3207 [25.01.2020].
43. Hochstetter, E., H. J. Greve, H. Gumin. *Herrn von Leibniz' Rechnung mit Null und Eins*, Siemens-Aktien-Ges. [Abt. Verlag], 1979.
44. Hubka, K. The late seventeenth-century lullism in Caspar Knittel's, *Collectanea Franciscana*, 51, 1981, pp. 65–82.
45. Idel, M. *Kabbalah: New Perspectives*, New Haven-London: Yale University Press, 1988.
46. Idel, M. *Language, Torah and Hermeneutics in Abraham Abulafia*, New York: State University of New York Press, 1988.

47. Ineichen, R. Leibniz, Caramuel, Harriot und das Dualsystem, *Mitteilungen der deutschen Mathematiker-Vereinigung*, 16(1), 2008, pp. 12–15.
48. Ioannis Caramuelis. *Mathesis biceps vetus et nova*, Officinâ Episcopali, 1670; https://books.google.pl/books?id=KRtetV1MJnkC&printsec=frontcover&source=gbs_book_other_versions_r&redir_esc=y#v=onepage&q&f=false [25.01.2020].
49. Izquierdo, S. *Pharus scientiarum ubi quidquid ad cognitionem humanam humanitùs acquisibilem pertinet, ubertim iuxtà, atque succinctè pertractaur*, Lugduni: Claudii Bovrgeat, Mich. Lietard, 1659; https://www.europeana.eu/portal/en/record/9200110/BibliographicResource_1000126649867.html [25.01.2020].
50. Johannes de Fontana, *Bellicorum instrumentorum liber cum figuris* – BSB Cod.icon. 242, Venedig, 1420 – 1430, https://daten.digitale-sammlungen.de/~db/0001/bsb00013084/images/index.html [16.01.2020].
51. John Gorman, M., and N. Wilding, *The Correspondence of Athanasius Kircher. The World of a Seventeenth Century Jesuit*, https://archimede.imss.fi.it/kircher/index.html [16.01.2020].
52. Kasik, S. *The Esoteric Codex: Christian Kabbalah*, Morrisville, North Carolina: Lulu.com, 2015.
53. Kennedy, H. *Peano: Life and Works of Giuseppe Peano*, Hubert Kennedy, 2006.
54. Kircher, A. *Prodromus coputs sive aegyptiacus*, Romae, 1636; https://books.google.co.th/books?id=KnATAAAAQAAJ&printsec=frontcover&hl=th&source=gbs_ge_summary_r&cad=0#v=onepage&q&f=false [25.01.2020].
55. Kircher, A. *Magnes, sive de arte magnetic*, Romae: Hermanni Scheus sub signo Reginae, 1641; https://archive.org/details/bub_gb_nK1DAAAAcAAJ/mode/2up [25.01.2020].
56. Kircher, A. *Lingua aegyptica restitute*, Romae: Hermani Scheus, apud Ludovicum Grignanum, 1643. https://books.google.pl/books?id=qEtB1x0frAIC&pg=PP21&dq=Lingua+aegyptica+restituta&hl=th&sa=X&ved=0ahUKEwiNgKHqxMLnAhXB66QKHVeTBNEQ6AEIYTAF#v=onepage&q=Lingua%20aegyptica%20restituta&f=false [25.01.2020].

57. Kircher, A. *Musurgia universalis, sive ars magna consoni et dissoni*, in x libros digesta, Vols. 1–2, Rome: Franceso Corbelletti, 1650; https://archive.org/details/bub_gb_97xCAAAAcAAJ/page/n8/mode/2up [25.01.2020].
58. Kircher, A. *Polygraphia nova et universalis, ex combinatoria arte detecta (etc.) in tria syntagmata distribute*, Romae: Varesius, 1663; https://books.google.co.th/books?id=-zVOAAAAcAAJ [25.01.2020].
59. Kircher, A. *Ars magna sciendi*, Amsterdam: Joannem Janssonius à Waesberge & Viduam Elizei Weyerstraet, 1669; https://archive.org/search.php?query=ars%20magna%20sciendi [25.01.2020].
60. Kircher, A. *Mundus subterraneus*, Amsterdam: Joannem Janssonius à Waesberge & Filios, 1678; https://archive.org/details/mundussubterrane02kirc/page/n6/mode/2up [25.01.2020].
61. *Kircher, Athanasius*, New Catholic Encyclopedia SPAIN, E. T., https://www.encyclopedia.com/people/science-and-technology/mathematics-biographies/athanasius-kircher [16.01.2020].
62. *Kircher's Mathematical Organ*, https://archimede.imss.fi.it/kircher/emathem.html [02.02.2020].
63. Klotz, S. Ars combinatoria oder 'musik ohne Kopfzerbrechen' Kalküle des musikalischen von Kircher bis Kirnberger, *Musiktheorie*, 3, 1999, pp. 231–247.
64. Knittel, C. *Via regia ad omnes scientias et artes. Hoc est: Ars universalis scientiarum omnium artiumque arcana facilius penetrandi*, Praque: Typis Universitatis Carolo-Ferdinandeae Pragae, 1682; https://reader.digitale-sammlungen.de/resolve/display/bsb11110333.html [16.01.2020].
65. Knuth, D. E. *The art of computer programming*, Vol. IV Generating all trees. History of Combinatorial Generation, Boston: Addison-Wesley, 2006; https://openlibrary.org/works/OL14941576W/Art_of_Computer_Programming_Volume_4_Fascicle_4_The [16.01.2020].
66. Kopania, J. Leibniz i jego Bóg. Rozważania z Voltaire'em w tl, *Studia z Historii Filozofii*, 3(9), 2018, pp. 69–101.

67. Künzel, W. *The Birth of the MACHINE: Raymundus Lullus and His Invention*, http://www.c3.hu/scca/butterfly/Kunzel/synopsis.html [18.02.2020], http://catalog.c3.hu/index.php?page=work&id=328&lang=EN [18.02.2020].
68. Le Myésier, T. *Rimundus Lullus: Electorium parvum seu breviculum*, St. Peter Perg. 92 Karlsruhe: Dr Ludwig Reichert, 1989.
69. Leibniz, G. W. *Dissertatio de arte combinatorial*, Lipsiae: Joh. Simon. Fickium et Jolh. Polycarp, Senboldum, 1666; abirintoermetico.com/12ArsCombinatoria/Leibniz_G_W_Dissertatio_de_Arte_combinatoria.pdf [18.02.2020]; https://archive.org/details/ita-bnc-mag-00000844-001/page/n11/mode/2up [18.02.2020].
70. Leibniz, G. W. *De progressione dyadic*, Vol. Pars I, 1679.
71. Leibniz, G. W. *Brief an den Herzog von Braunschweig-Wolfenbüttel Rudolph August, 2. Januar 1697*, http://www.fh-augsburg.de/~harsch/germanica/Chronologie/17Jh/Leibniz/lei-bina.html [18.02.2020].
72. Leibniz, G. W. Explication de l'arithmétique binaire, qui se sert des seuls caractères 0 et 1, avec des remarques sur son utilité, et sur ce quélle donne le sens des anciennes figures Chinoises de Fohy, *Memoires de l'Académie Royale des Sciences*, 3, 1703, pp. 85–89.
73. Leibniz, G. W. *Gesammelte Werke. Aus den Handschriften der Königlichen Bibliothek zu Hannover*, G. H. Pertz, C. L. Grotefend, & C. I. Gerhardt (Eds.), Hannover: Hahnschen Hof-Buchhandlung, 1843; https://archive.org/details/bub_gb_u1AJAAAAQAAJ/page/n329/mode/2up [18.02.2020].
74. Leibniz, G. W. Dialogus, [in:] C. I. Gerhardt (Ed.), *Die philosophischen Schriften von G. W. Leibniz*, Vol. 7, Berlin, 1890, pp. 190–193.
75. Leibniz, G. W. *Philosophische Schriften*, Vol. 7; C. I. Gerhardt (Ed.), Berlin: Weidmann, 1890.
76. Leibniz, G. W. Machina arithmetica in qua non additio tantum et subtractio sed et multiplicatio nullo, divisio vero paene nullo animi labore peragantur, [in:] D. E. Smith (Ed.), *A source Book in*

Mathematics, 1st ed., New York: McGraw Hill Book Company, 1929, pp. 173–181; https://archive.org/details/sourcebookinmath00smit/mode/2up [18.02.2020].
77. Leibniz, G. W. Letters to Nicolas Remond, [in:] L. E. Loemker (Ed.), *Philosophical papers and letters*: A selection, 2nd ed., Vol. 2, Dordrecht: Springer, 1989, pp. 654–660.
78. Leibniz, G. W. Letter to Thomas Hobbes, [in:] L. E. Loemker (Ed.), *Philosophical papers and letters*, Vol. 2, Dordrecht: Springer, 1989.
79. Leibniz, G. W. *Leibniz korrespondiert mit China: der Briefwechsel mit den Jesuitenmissionaren (1689–1714)*. Frankfurt am Main: V. Klosterman, 1990.
80. Leibniz, G. W. *Sämtliche Schriften und Briefe*, Vol. 1; Herausgegeben von der Berlin-Brandenburgischen Akademie der Wissenschaften und der Akademie der Wissenschaften in Göttingen, Berlin: Akademie Verlag, 2006.
81. Leinkauf, T. Mundus combinatus und ars combinatoria als geistesgeschichtlicher Hintergrund des Muesum Kircherianum in Rom, [in:] A. Grote (Ed.), *Macrocosmos in Microcosmo. Die Welt in der Stube. Zur Geschichte des Sammelns 1450 bis 1800*, Oplanden, 1994, pp. 535–553.
82. Leinkauf, T. Athanasius Kircher, [in:] W. S.-B. Helmut Holzhey Friedrich Ueberweg (Ed.), *Grundriss der Geschichte der Philosophie. Die Philosophie des 17. Jahrhunderts*, Vols. 4, Das Heilige Romische Reich Deutscher Nation, Nord- und Ostmitteleuropa, Basel: Schwabe Verlagsgruppe AG Schwabe Verlag, 2001, pp. 269–290.
83. Leinkauf, T. Lullismus, [in:] W. S.-B. Helmut Holzhey Friedrich Ueberweg (Ed.), *Grundriss der Geschichte der Philosophie. Die Philosophie des 17. Jahrhunderts*, Vols. 4, Das Heilige Römische Reich Deutscher Nation, Nord- und Ostmitteleuropa, Basel: Schwabe Verlagsgruppe AG Schwabe Verlag, 2001, pp. 235–269.
84. Ligonnière, R. *Prehistoria i historia komputerów*, Wrocław: Ossolineum, 1992.
85. Link, D. Scrambling T-R-U-T-H. Rotating letters as a material form of thought, [in:] S. Zielinski & E. Fuerlus (Eds.), *Variantology*, Vols. 4. On Deep Time Relations of Arts, Sciences and Technologies in the Arabic-Islamic World and Beyond, Cologne: König, 2010, pp. 215–266;

http://www.alpha60.de/research/scrambling_truth/ [02.02.2020].
86. Lohr, C. Christianus arabicus, cuius nomen Raimundus Lullus, *Freiburger Zeitschrift für Philosophie und Theologie*, 31(1–2), 1984, pp. 57–88.
87. Lull, R. Ars brevis, [in:] A. Bonner (Ed.), *Selected works of Ramon Llull (1232–1316)*, Vol. 1, Princeton, N. J.: Princeton University Press, 1985, pp. 579–646.
88. Lull, R. *The Ultimate General Art*, Y. Dambergs (Ed.), Quebec, 2009;
https://academiaanalitica.files.wordpress.com/2016/10/lullus-ars-generalis-ultima-ars-magna.pdf [02.02.2020].
89. Lull, R. *Selected works of Ramon Llull (1232–1316)*, Vols. 1–2, A. Bonner (Ed.), Princeton: Princeton University Press, 1985.
90. Lullus, R. *Raymundi Lullij doctoris illuminati de nuoa logica,* 1512;
https://books.google.pl/books?id=WR1XAAAAcAAJ&pg=PT278&lpg=PT278&dq=logica+now+lullus&source=bl&ots=DWiNFeEBrE&sig=PgyeyPJlAmsTEKwqd4TQBHiUY3Y&hl=en&sa=X&ved=0ahUKEwirxIK3lYDRAhXFWxQKHWfLB_c4ChDoAQhFMAg#v=onepage&q=logica%20now%20lullus&f=false [02.02.2020].
91. Lullus, R. *Ars generalis ultima*, Frankfurt/Main: Minerva, 1970.
92. Malcolm, N., and J. Stedall. *John Pell (1611–1685) and His Correspondence with Sir Charles Cavendish: The Mental World of an Early Modern Mathematician*, Oxford: Oxford University Press, 2005.
93. McCarthy, J., M. L. Minsky, N. Rochester, C.E. Shannon. *A proposal for the dartmouth summer research project on artificial intelligence, August 31, 1955*,
http://www-formal.stanford.edu/jmc/history/dartmouth/dartmouth.html [02.02.2020].
94. McCorduck, P. *Machines who think: A personal inquiry into the history and prospects of artificial intelligence*, 2nd ed., Natick, MA: A. K. Peters, 2004.
95. Mindel, N. *The Maharal to the rescue: And other stories of rabbi Yehuda Loew of Prague*, New York: Merkos Publications, 2007.
96. Miniati, M. Les „cistae mathematicae" et l'organisation des connaissances au xviie siècle, [in:] C. Blondel, F. Parot, A.

Turner, M. Williams (Eds.), *Studies in the History of Scientific Instruments*, London: Turner Books, 1989, pp. 43–51.
97. Minsky, M. *Computation: Finite and Infinite Machines*, New York: Engelwood Cliffs, Prentice-Hall, 1967.
98. Minsky, M. *The Emotion Machine. Commonsense Thinking, Artificial Intelligence, and the Future of the Human Mind*, New York: Simon & Schuster, 2006.
99. Nicholas of Cusa, De coniecturis (On surmises), [in:] J. Hopkins (Ed.), *Nicholas of Cusa: Metaphysical speculations*, Vol. 2, Minneapolis: The Arthur J. Banning Press, 2000, pp. 163–297; https://jasper-hopkins.info/DeConi12-2000.pdf [25.01.2020].
100. Peano, G. De latino sine flexione: Lingua auxiliare internationale, *Revue de Mathématiques*, 8, 1903, pp. 74–83; http://www.gutenberg.org/files/35803/35803-h/35803-h.htm [25.01.2020].
101. Peckhaus, V. Leibniz als Identifikationsfigur der britischen Logiker des 19. Jahrhunderts, [in:] *VI. Internationaler Leibniz-Kongreß. Vorträge. I. Teil*, Hannover, 18.–22.7.1994, Hannover: Gottfried-Wilhelm-Leibniz-Gesellschaft, 1994, pp. 589–596.
102. Peckhaus, V. *Logik, Mathesis universalis und allgemeine Wissenschaft. Leibniz und die Wiederentdeckung der formalen Logik im 19. Jahrhundert*, Berlin: Akademie Verlag, 1997.
103. Peckhaus, V. 19th century logic between philosophy and mathematics, *The Bulletin of Symbolic Logic*, 5(4), 1999, pp. 433–450.
104. Peckhaus, V. Leibniz und die britischen Logiker des 19. Jahrhunderts, *Erweiterte und revidierte Fassung des Vortrag, gehalten am 22. Juli 1994 auf dem VI. Internationalen Leibniz-Kongreß in Hannover*, 1994.
105. Peckhaus, V. Leibniz's influence on 19th century logic, [in:] E. N. Zalta (Ed.), *The Stanford encyclopedia of philosophy*, Metaphysics Research Lab, Stanford University, 2018; https://plato.stanford.edu/archives/win2018/entries/leibniz-logic-influence/ [25.01.2020].
106. Priani, E. Ramon llull, [in:] E. N. Zalta (Ed.), *The Stanford encyclopedia of philosophy*, Metaphysics Research Lab, Stanford University, 2017; https://plato.stanford.edu/archives/spr2017/entries/llull/ [25.01.2020].
107. *Ramon Llull*, World Heritage Encyclopedia,

http://self.gutenberg.org/article/WHEBN0000069677/Ramon%20Llull [25.01.2020].
108. Ratzabi, H. *What Is Gematria? Hebrew numerology, and the secrets of the Torah*, https://www.myjewishlearning.com/article/gematria/ [25.01.2020].
109. Reilly, C. Father A. Kircher, S. J.: Master of an hundred arts, *Studies: An Irish Quarterly Review*, 44(176), 1955, pp. 457–468.
110. Reilly, C. *Athanasius Kircher S. J.: Master of a Hundred Arts*, 1602–1680, Vol. 1, Wiesbaden: Edizioni del Mondo, 1974.
111. *Researchers and Philosophers Write about Kabbalah*, http://www.kabbalah.info/eng/content/view/frame/80159?/eng/content/view/full/80159&main [25.01.2020].
112. Rheagan, M. *Decoding the Medieval Volvelle. Made from circles of paper or parchment, the volvelle was part timepiece, part floppy disk, and part crystal ball*, http://blogs.getty.edu/iris/decoding-the-medieval-volvelle/ [25.01.2020].
113. Scholz, H. *Geschichte der Logik*, Berlin: Junker und Dünnhaupt, 1931.
114. Schott, G. *Ioco-seriorum naturae et artis, siue, Magiae naturalis centuriae tres: accessit Diattibe de prodigiosis crucibus*, Würzburg, 1666; https://searchworks.stanford.edu/view/209930 [25.01.2020].
115. Schuhmann, K. Leibniz's letters to Hobbes, *Studia Leibnitiana*, 37(2), 2005, pp. 147–160.
116. Serra, Y. *Leibniz's De Progressione Dyadica Manuscript*, http://www.bibnum.education.fr/sites/default/files/69-analysis-leibniz.pdf [20.01.2020].
117. Shelley, M. *Frenkenstein; or, the Modern Prometheus*, Vols. 1–3, London: Lackington, Hughes, Harding, Mavor, & Jones, Finsbury Square, 1818; http://www.gutenberg.org/ebooks/41445 [20.01.2020].
118. Shirley, J. W. Binary numeration before Leibniz, *American Journal of Physics*, 19(8), 1951, pp. 452–454.
119. Stolzenberg, D. (Ed.), *The Great Art of Knowing: The Baroque Encyclopedia of Athanasius Kircher*, Stanford: Stanford University Libraries, 2001.
120. Swetz, F. J. Leibniz, the Yijing, and the religious conversion of the Chinese, *Mathematics Magazine*, 76(4), 2003, pp. 276–291.

121. Szapiro, G. G. *Numbers rule: The vexing mathematics of democracy, from Plato to the present*, Princeton, New Jersey: Princeton University Press, 2010.
122. Tenge-Wolf, V. *Ramon Llull im WWW – Projektbeschreibung*, http://www.theol.uni-freiburg.de/disciplinae/dqtm/forschung/raimundus-lullus [02.02.2020].
123. *The Correspondence of Athanasius Kircher (2,741 letters)*, http://emlo-portal.bodleian.ox.ac.uk/collections/?catalogue=athanasius-kircher [25.01.2020].
124. *The Ternary Calculating Machine of Thomas Fowler*, http://mortati.com/glusker/fowler/fowlerbio.htm [25.01.2020].
125. Trzęsicki, K. From the idea of decidability to the number Ω, *Studies in Grammar, Logic and Rhetoric*, 9(22), 2006, pp. 73–142; http://logika.uwb.edu.pl/studies [25.01.2020].
126. Trzęsicki, K. Leibniza idea systemu binarnego, [in:] J. Kopania & H. Święczkowska (Eds.), *Filozofia i myśl społeczna XVII w.*, Białystok, 2006, pp. 183–203.
127. Trzęsicki, K. Leibnizjańskie inspiracje informatyki, *Filozofia Nauki*, 55(3), 2006, pp. 21–48.
128. Turing, A. M. Computing machinery and intelligence, *Mind*, 49(236), 1950, pp. 433–460; www.csee.umbc.edu/courses/471/papers/turing.pdf [25.01.2020].
129. Urvoy, D. *Penser l'Islam. Les présupposés Islamiques de l'"art" de Lull*, Paris: J. Vrin, 1980.
130. van Heijenoort, J. (Ed.), *From Frege to Gödel. A Source Book in Mathematical Logic 1879—1931*, 1st ed., Cambridge Mass.: Harvard University Press, 1967.
131. Vass, P. *The Power of Three: Thomas Fowler. Devon's Forgotten Genius*, Boundstone Books, 2016.
132. Verburg, P. A. Hobbes' calculus of words, [in:] *International Conference on Computational Linguistics COLING 1969*: Preprint no. 39. Sånga Säby, Sweden, 1969; https://www.aclweb.org/anthology/C69-3901 [25.01.2020].
133. Vincent, J. *Report: OpenAI's new multitalented AI writes, translates, and slanders. A step forward in AI text-generation that also spells trouble*, https://www.theverge.com/2019/2/14/18224704/ai-machine-learning-language-models-read-write-openai-gpt2 [25.01.2020].

134. Waddell, M. A. Magic and artificie in the collection of Athanasius Kircher, *Endeavour*, 34(1), 2010, pp. 30–34.
135. Watkins, R. P. *Computer Problem Solving*, Sydney: John Willey & Sons Australasia, 1974.
136. Wilkins, J. *An Essay towards a Real Character, and a Philosophical Language*, London, 1668;
https://archive.org/details/AnEssayTowardsARealCharacterAndAPhilosophicalLanguage/mode/2up [25.01.2020].
137. Wilkins, J. *Mercury, or, The Secret and Swift Messenger: Shewing, How a Man May with Privacy and Speed Communicate His Thoughts to a Friend at any Distance*, London, 1694;
https://archive.org/details/gu_mercuryorthes00wilk/page/n10/mode/2up [25.01.2020].

Notes

1. See [133]. However, GPT-2 is not made publicly available due to possible abuses. An example of text written by GPT-2 can be seen here: https://lionbridge.ai/articles/this-entire-article-was-written-by-an-ai-open-ai-gpt2/ [02.02.2020].
2. Please see the conference materials: July 13-15, 2005, the *Dartmouth Artificial Intelligence Conference: The Next Fifty Years*, https://www.dartmouth.edu/~ai50/homepage.html [20.01.2020].
3. Base de Dades Ramon Llull (Llull DB), please see:
http://www.ub.edu/llulldb/index.asp?lang=ca [02.02.2020].
4. The web page of: https://www.csic.es/es/el-csic [02.02.2020], please see also [3].
5. "The Creation, or language, is an adequate subject of the science of Kabbalah […] That is why it is becoming clear that its wisdom governs the rest of the sciences. Sciences such as theology, philosophy and mathematics receive their principles and roots from it. And therefore these sciences (scientiae) are subordinate to that wisdom (sapientia); and their [= the sciences] principles and rules are subordinate to it [= the Kabbalah] principles and rules; and therefore their [= the sciences] mode of argumentation is insufficient without it [= the Kabbalah]" [111].
6. The full sentence is: "Words are wise men's counters, they do but reckon by them: but they are the money of fools, that value them by the authority of an Aristotle, a Cicero, or a Thomas, or any other doctor whatsoever, if but a man" [42].

7. The works by Athanasius Kircher (1602 – 1680) are available here: https://archive.org/search.php?query=creator%3A"Kircher%2C+Athanasius%2C+1602-1680" [02.02.2020].
8. For the web page of the Museum of Jurassic Technology, please see: http://mjt.org/ [02.02.2020].
9. The web page of the Kircher's Museum: https://archimede.imss.fi.it/kircher/emuseum.html [25.01.2020]. Kircher's ethnographic collection is located in Rome, Pigorini, the National Museum of Prehistory and Ethnography.
10. The web page of Museo Galileo: https://www.museogalileo.it//en [02.02.2020].
11. A student of Athanasius Kircher, Gaspar Schott, publishes a treatises on the wonders of scientific innovation, please see: http://www.rarebookroom.org/pdfDescriptions/schioc.pdf [02.02.2020].
12. An appropriate phrase in Latin: *Petrus noster amicus, venit ad nos qui portavit tuas litteras ex quibus intellexi tuum animum atque faciam iuxta tuam voluntatem.*
13. The model of Leibniz-inspired binary machine was built in the years of 2003–2004 by E. Stein and G. Weber, Das Institut für Baumechanik und Numerische Mechanik, Leibniz Universität Hannover.
14. The ternary calculating machine of Thomas Fowler, please see: http://mortati.com/glusker/fowler/fowlerbio.htm [02.02.2020].
15. Similar statements are contained in other texts of the quoted volume, e.g. on pages [75, pp. 26, 64-65, 125].
16. Name is formed from 'LIST Processor.'
17. The web page of Leibniz-Sozietät der Wissenschaften zu Berlin e.V., please see: http://www.leibnizsozietaet.de/ [02.02.2020].
18. For indicating the original texts of Descartes and their translation, and also additional data, I thank Jerzy Kopani.
19The full note is as follows: "*Excogidad in curru inter Hanoveram et Peinam* 14. October. 1895 G. L. R. *Machina arithmetica cum verbis* SUPRA HOMINEM. [*Nam hominem maximorum calculorum et promtitudine et securitate vincit.*] *Miramur ratio est divina quod indita rebus*: S u p r a h o m i n e m *humana est machina facta manu. Quanta Deum fecisse putas hominem super? Ecce* S u p r a h o m i n e m humana *est machina facta manu.*"

Conjunctive and Disjunctive Limits: Abstract Logics and Modal Operators

Alexandre Costa-Leite

University of Brasilia
Campus Darcy Ribeiro – ICC – Ala Norte
CEP 70910-900
Brasilia DF Brazil
e-mail: costaleite@unb.br

Edelcio G. de Souza

University of Sao Paulo
Rua Luciano Gualberto, 315
CEP 05508-010
Sao Paulo Brazil
e-mail: edelcio.souza@usp.br

1. Introduction

There are families of concepts organized by some order and some kind of hierarchy. This phenomenon occurs in distinct areas of logic: sequences of sentences can be systematized to highlight the most essential element in the sequence (the sovereign object in the hierarchy). In this article, we use the concept of limit of a given sequence to redefine the notions of conjunctive limit and disjunctive limit in the universe of abstract logic.[1] By means of this strategy, we can formulate specific standards of logical possibility as well logical necessity pointing out that the same procedure could be extended to a great variety of sequences of objects (with very different natures, indeed).

We start introducing main useful concepts from abstract logic and, then, in section 3, we present some notions such as those of *conjunctive limit, disjunctive limit* as well the concepts of *conjunctive logic* and *disjunctive logic*. In section 4, some of these ideas are applied to reason about levels of modal operators.

2. Concepts in Abstract Logics

We establish several basic preliminary and standard concepts in the realm of general abstract logic following initial ideas developed by Alfred Tarski in [9] and [10].[2] This approach to logical consequence sounds awesome and very useful allowing us to be well oriented through the incredible plurality of rationalities displayed by the great variety of logical systems.

An *abstract logic* is a pair $L = (S, Cn_L)$ such that S is a non-empty set and Cn_L is a map

$$Cn_L : \wp(S) \to \wp(S)$$

in the power set of S. The operator Cn_L should satisfy:[3]

i. *Inclusion*: $A \subseteq Cn(A)$.
ii. *Idempotency*: $Cn(Cn(A)) = Cn(A)$.
iii. *Monotonicity*: $Cn(A) \subseteq Cn(A \cup B)$.

We call S the *domain* or *universe* of L and Cn_L is its *consequence operator*. Elements of S are called *sentences* and, therefore, we are concerned with logical consequence defined for sentences.[4]

As it is well-known, consequence operators are connected with consequence relations by means of a very natural relationship. Given an abstract logic $L = (S, Cn_L)$, it is feasible to define a binary relation

$$\vdash_L \subseteq \wp(S) \times S$$

such that:

$$A \vdash_L a \text{ iff (if and only if) } a \in Cn_L(A).$$

We call $\vdash_L$ the *consequence relation* of L. It is easy to see that $\vdash_L$ satisfies:[5]

I. *Inclusion*: If a ∈ A, then A ⊢ a.
II. *Transitivity*: If B ⊢$_L$ a and A ⊢ b for all b ∈ B then A ⊢ a.
III. *Monotonicity*: If A ⊢ a and A ⊆ B, then B ⊢ a.

I and III are immediate. For II, suppose that a ∈ Cn(B) and b ∈ Cn(A) for all b ∈ B. Thus, B ⊆ Cn(A). By idempotency and monotonicity, we have:

$$Cn(B) \subseteq Cn(Cn(A)) = Cn(A).$$

Then, a ∈ Cn(A), that is, A ⊢ a.

Now, let L = (S, Cn$_L$) be an abstract logic. We say that:

(a) A ⊆ S is *L-limited* iff Cn$_L$(A) ≠ S.[6] Otherwise, A is *L-unlimited*;
(b) A sentence c ∈ S is a *0-sentence* iff Cn$_L$({c}) is L-unlimited. Moreover, if t ∈ Cn$_L$(∅) we say that t is a *1-sentence*.

We denote by **1**$_L$ and **0**$_L$ the set of all 1-sentences and 0-sentences, respectively. The four above notions are in some sense related with the traditional concepts of consistency, inconsistency, contradiction and tautology, respectively. Here these concepts are dressed with new clothes to be more adaptable to our purposes departing from a general perspective.

Let L = (S, Cn$_L$) be an abstract logic. We can use the consequence operator Cn$_L$ in order to define a partial order in the set of sentences S.
If x, y ∈ S, we define:

$$x \leq y \text{ iff } \{x\} \vdash_L y.$$

It is clear that ≤ is reflexive and transitive. Then, (X, ≤) is a partially ordered set. In this setting, we can take into account upper and lower bounds, supremum, infimum, maximal and minimal elements etc.

We use Cn$_L$ to define, for x, y ∈ S, an equivalence relation between elements of S in the following way:

$$x \equiv y \text{ iff } Cn_L(\{x\}) = Cn_L(\{y\}).$$

In this case, we have: $\{x\} \vdash_L y$ and $\{y\} \vdash_L x$. It is easy to see that $\equiv$ is an equivalence relation and, as usual, we have that:

$$[x]_\equiv = \{y \in S : x \equiv y\}.$$

Therefore, the quotient set is given by

$$S/_\equiv = \{[x]_\equiv \in \wp(S) : x \in S\}.$$

The order relation $\leq$ is transferred to the set $S/_\equiv$:[7]

$$[x] \leq [y] \text{ iff } x \leq y.$$

This construction does not depend on the representatives x and y.

Given an abstract logic $L = (S, Cn_L)$, if the sets $\mathbf{1}_L$ and $\mathbf{0}_L$ are not empty, then the sets $\mathbf{1}_L$ and $\mathbf{0}_L$ are the greatest and the lowest elements in the ordered set $S/_\equiv$.

Dealing with logics from this abstract viewpoint sounds very elegant and useful, especially considering the mess caused by the plurality of rationalities that one can find in the market. And, more important, this approach is essential to our next definitions.

3. Conjunctive and Disjunctive Limits

The original ideas of *conjunctive* and *disjunctive* limits introduced in this section appeared inside a different framework in [4]. These concepts are here reformulated in the spirit of abstract logic. From now on, it follows the main contributions of this paper.

Consider an abstract logic $L = (S, Cn_L)$. Let $(x_i)_{i \in \omega}$ be a sequence of elements of S. We say that $[x]_\equiv \in S/_\equiv$ is a **conjunctive limit** of $(x_i)_{i \in \omega}$ iff there exists $k \in \omega$ such that for $i \geq k$, we have $\{x\} \vdash_L x_i$ (or, that is the same, $x \leq x_i$). The set of all conjunctive limits of $(x_i)_{i \in \omega}$ is denoted by $\mathbf{LIM}^c(x_i)$. Notice that if c is a 0-sentence, then $[c]_\equiv = \mathbf{0}_L$ is a conjunctive limit of all sequences of elements of S. This allows us to talk about a formula from which all other formulas can be derived.[8]

The construction above can be dualized. We say that $[x]_\equiv \in S/_\equiv$ is a **disjunctive limit** of $(x_i)_{i \in \omega}$ iff there exists $k \in \omega$ such that for $i \geq k$, we have $\{x_i\} \vdash_L x$ (or, that is the same, $x_i \leq x$). The set of all disjunctive limits of $(x_i)_{i\in\omega}$ is denoted by $\mathbf{LIM}^d(x_i)$. In this case, if t is a 1-sentence, then $[t]_\equiv = \mathbf{1}_L$ is a disjunctive limit of all sequences of elements of S. Now, this allows us to talk about a formula which is a consequence of all other formulas.

An abstract logic $L = (S, Cn_L)$ is **conjunctive** iff for all sequence $(x_i)_{i\in\omega}$ of elements of S, the set $\mathbf{LIM}^c(x_i)$ has the minimum. In this case, we define:

$$\lim{}^c(x_i) := \min(\mathbf{LIM}^c(x_i)).$$

A logic $L = (S, Cn_L)$ is **properly conjunctive** iff for all sequence $(x_i)_{i\in\omega}$ of elements of S, the set $\mathbf{LIM}^c(x_i)$ has the minimum that it is not $\mathbf{0}_L$.

We also have an immediate dual concept: an abstract logic $L = (S, Cn_L)$ is **disjunctive** iff for all sequence $(x_i)_{i\in\omega}$ of elements of S, the set $\mathbf{LIM}^d(x_i)$ has the maximum. In this case, we define:

$$\lim{}^d(x_i) := \max(\mathbf{LIM}^d(x_i)).$$

A logic $L = (S, Cn_L)$ is **properly disjunctive** iff for all sequence $(x_i)_{i\in\omega}$ of elements of S, the set $\mathbf{LIM}^d(x_i)$ has the maximum that it is not $\mathbf{1}_L$.

On one hand, if we consider classical propositional logic C, it is easy to see that C is a conjunctive and disjunctive logic. But C is neither properly conjunctive nor properly disjunctive. For example, the sequence of all propositional variables has no conjunctive and disjunctive limits different from $\mathbf{1}_L$ and $\mathbf{0}_L$. On the other hand, if we consider infinitary classical propositional logic C_∞, with infinitary conjunctions and disjunctions, we have a properly conjunctive and disjunctive logic.

The concept of a *finitely trivializable system* is used to refer to a logic containing a formula from which everything (i.e. all formulas in the language) can be deduced (cf. [6]). We can say that if a logic is conjunctive (but not properly conjunctive), then it is finitely trivializable. In this sense, the system C_ω in da Costa's hierarchy is not a conjunctive logic while C_1 is a conjunctive and a disjunctive logic.

4. Limits of Sequences and Modal Operators

We have argued (with Hilan Bensusan) in [1] that logical possibility and logical necessity are never absolute in the precise sense that what is logically possible in a given logic could be logically impossible in a different logic and vice-versa. The same applies, then, for logical necessity and, in more general terms, for all logical truths. So, it is inside a given logic that something is logically possible or not. We take, then, logical possibility with respect to a given logic as the largest concept in such a way that all kinds of empirical possibility (weaker possibilities) are particular cases of it (let's call them X-possibilities for X being a particular theory, as suggested in [3]). In this way, if something is X-possible, then it is logically possible (in a formal system taken as underlying logic of a given theory).[9] This obviously gives a clue to the fact that logical possibility is a kind of limit of a sequence of modal ◊-formulas.[10] Conversely, logical necessity can also be viewed as a sort of limit of a sequence of □-formulas, considering that if something is a logical necessity, then it is an X-necessity. So, for this reason, let us concentrate here in the case of modal operators, especially those of the form ◊ and of the form □.

Assume a family of normal modal logics with finitely many modal operators. Let $Y_1,...,Y_n$ be this multimodal system such that for each Y_i there is a $◊_i$ and a respective definable $□_i$. From the viewpoint of abstract logic, this system is a multimodal abstract logic (S,Cn) such that S contains sequences of modal operators $\{◊_i\}_{i \in \omega}$ and $\{□_i\}_{i \in \omega}$. As mentioned, these operators could represent different kinds, degrees, levels of possibilities and necessities (X-possibilities and so on). Moreover, suppose that (S,Cn) is a properly conjunctive and a properly disjunctive logic.

Let x be an element of S and consider the sequence $\{◊_i x\}_{i \in \omega}$ of elements of S. In this way, we could define logical possibility ◊ in (S,Cn) as a disjunctive limit of this sequence, that is:

$$◊x := \lim^d(◊_i x) = \max(\mathbf{LIM}^d(◊_i x)).$$

Similarly, we could define the logical necessity □ in (S,Cn) as a conjunctive limit of this sequence, that is:

$$□x := \lim^c(□_i x) = \min(\mathbf{LIM}^c(□_i x)).$$

If we take logical possibility and logical necessity as above, we could say that multimodal systems with interactive axioms regulating levels of modal operators can be viewed as logics in which there are conjunctive and disjunctive limits. So, in the first case, we would have a logic (where $\otimes$ represents fusions of logics)[11]

$$Y_1 \otimes \ldots \otimes Y_n \otimes (\Diamond_1 a \to \Diamond_2 a) \otimes \ldots \otimes (\Diamond_{n-1} a \to \Diamond_n a)$$

and logical possibility $\Diamond a$ is $\Diamond_n a$. The relevant fact is that all other kinds of possibility imply logical possibility in such a way that this one can be viewed, therefore, as a disjunctive limit. In the second case, conversely, we would have a logic

$$Y_1 \otimes \ldots \otimes Y_n \otimes (\Box_n a \to \Box_{n-1} a) \otimes \ldots \otimes (\Box_2 a \to \Box_1 a)$$

and logical necessity $\Box a$ is $\Box_n a$. Now, the relevant fact is that logical necessity implies all kinds of necessity and, therefore, it is a conjunctive limit. So, in multimodal logics with ordered modal operators, it is very natural to think about conjunctive and disjunctive limits. Thus, we can say that logical possibility is the disjunctive limit of a sequence of weaker sorts of possibilities (as each X-possibility implies logical possibility) and the dual logical necessity is the conjunctive limit of a sequence of stronger kinds of necessities (if something is a logical necessity, then it is X-necessary).

Considering that logical possibility (and its dual logical necessity) are always determined with respect to a given logic, it follows that hierarchies of weaker possibilities (and necessities) are also with respect to a given logic. Therefore, for each logical diamond or box, we have a respective hierarchy of X-possibilities (necessities) in a previous theory.

5. Conclusion

Limits of sequences of formulas (and, in particular, modal formulas) have a wide variety of applications. Treating logical possibility and logical necessity as disjunctive and conjunctive limits suggests that it is feasible to define other dual concepts in a similar fashion. The notion of *disjunctive limit* of a sequence involves the idea that the disjunctive limit can be derived from any element in the sequence, and disjunctive limit can be derived from any element in the sequence, and it allows us to define the notion of *disjunctive logic*. The idea of *conjunctive limit* of a sequence accepts that a conjunctive limit implies any element of the sequence, and, as such, it can be used to define a *conjunctive logic*. The contributions of this paper are conceptual in the sense that definitions were designed to be applied in logical research. As in many situations we find hierarchies of sentences, limits can always be launched, and, therefore, definitions suggested here have a large scope of applications. It seems that these abstractions also facilitate attempts to model some situations in mathematics and philosophy.

Acknowledgement

This paper is dedicated to Jan Woleński for the celebration of his 80th birthday. Thanks to Andrew Schumann for the invitation to contribute with this special issue of *Studia Humana*.

References

1. Bensusan, H., A. Costa-Leite, and E. G. de Souza. Logics and their galaxies, In A. Koslow and A. Buchsbaum (eds.), *The Road to Universal Logic*, vol. 2, Basel: Birkhäuser, 2015, pp. 243-252.
2. Beziau, J-Y. *Recherches sur la logique universelle (excessivité, négation et séquents*, PhD Thesis, Université Denis-Diderot Paris 7, 1995.
3. Costa-Leite, A. Logical properties of imagination. *Abstracta* 6 (1), 2010, pp. 103-116.
4. Costa-Leite, A., and E. G. de Souza. Implications and limits of sequences, *Studia Humana* 6 (1), 2017, pp. 18-24.
5. Costa-Leite, A. *Paraconsistência, modalidades e cognoscibilidade*, Available on PhilArchive: retrieved June 17, 2020 from https: //philarchive.org/archive/COSPME-5, 2019 (in Portuguese).
6. da Costa, N. On the theory of inconsistent formal systems, *Notre Dame Journal of Formal Logic* 15 (5), 1974, pp. 497-510.
7. Dunn, M, and G. Hardegree. *Algebraic Methods in Philosophical Logic*, New York: Oxford University Press, 2001.
8. Gabbay, D., A. Kurucz, F. Wolter, and M. Zakharyaschev. *Many-dimensional modal logics: theory and applications*, Studies in Logic and the Foundations of Mathematics, Amsterdam: Elsevier, 2003.

9. Tarski, A. On some fundamental concepts of metamathematics, In J. Corcoran (ed.), *Logic, Semantic, Metamathematics*, Second Edition, Indianapolis: Hackett Publishing Company, 1983.

10. Tarski, A. Fundamental concepts of the methodology of deductive sciences, In J. Corcoran (ed.), *Logic, Semantic, Metamathematics*, Second Edition, Indianapolis: Hackett Publishing Company, 1983.

Notes

1. These two concepts were initially proposed in [4], but without appeal to abstract logic.
2. A textbook relating ideas of algebra with logic in the domains of algebraic logic and algebra of logic can be found in [7].
3. We omit the subscript L.
4. Other forms of logical consequence could be defined taking into account objects without making any reference to the linguistic dimension of S.
5. Again, the subscript L is omitted.
6. This terminology is due to Jean-Yves Béziau in [2].
7. The subscript $\equiv$ is omitted.
8. In classical logic, a contradiction has this role, though it is not like this in all formal systems.
9. A hierarchy of diamonds have been used in [3] to build a combined logic of imagination, for instance.
10. A previous characterization of diamonds and boxes as limits of sequences of modal operators has been formerly developed in [5].
11. Cf. [8] for a roadmap with respect to combining logics in the environment of modal logics.

A Judgmental Reconstruction of Some of Professor Woleński's Logical and Philosophical Writings

Fabien Schang

Federal University of Goiás
Avenida Esperança, SN, Campus Samambaia - Conj. Itatiaia
Goiânia, 74690-900, Brasil
e-mail: schangfabien@gmail.com

1. Introduction: Neither Frege, Nor Suszko (Therefore Łukasiewicz?)

Suszko is known both for his eponymous acceptance of the 'Suszko Thesis', under which all logical systems whose consequence operator satisfies the criterion of structurality (or of extensionality) are bivalent systems, and for his rejection of the 'Frege's Axiom' (FA). We are going to focus on the latter, and more specifically on the different ways of opposing it. Suszko [27] is in some way opposed FA, which consists of two sub-propositions $(FA_1) - (FA_2)$:

FA_1 The referent of a sentence is its truth-value.
FA_2 This truth-value is either the True or the False.

Suszko rejects FA_1 and accepts FA_2. According to him, sentences do not express truth-values but situations, and this explains why Suszko distinguishes identity from material equivalence or biconditional since the rejection of FA_1 implies that two sentences may have the same truth-value without being identical. But there is another way to reject FA, by reasoning in reverse to Suszko and accepting FA_1 while rejecting FA_2. It is this position that we will associate with the name of Łukasiewicz and defend in this article, while seeking to justify it through several writings of Professor Woleński. It will thus be a matter of defending one direction of Polish logic against another, namely: many-valued logics of Łukasiewicz, as opposed to the 'non-Fregean' logic of Suszko.

Starting from a preliminary reflection on the meaning of the Principle of Contradiction (PC) and its analysis by Łukasiewicz, we will begin by distinguishing 'sentences' from 'propositions' and endorsing the many-valuedness entailed by the rejection of FA_2 through a general logic of *judgments*. Then we will review a certain number of logical questions treated by Jan Woleński in the light of this logic of judgments: the Principle of Bivalence (PB), and its various definitions; the relationship between logic and scepticism, and the concept of duality; the relationship between coherence and truth, Tarski's T-scheme, and the relativity of the concept of truth; negation, and the philosophical notion of 'nothingness'. We hope that the logical framework resulting from our non-Suszkian rejection of FA will confirm and clarify certain reflections of Professor Woleński on all these matters. Last but not least, we will insist on a formal tool essential to metalogical reflection and which Woleński frequently uses in the articles treated here: the *theory of opposition*.

2. Frege's Axiom and its Opponents

FA is neither true nor false strictly speaking, so neither are what Suszko and Łukasiewicz said about it. It is rather necessary to think of this metalogical axiom in terms of explanatory virtue: which position with regard to FA is the most insightful, from an explanatory point of view?

FA relates to PB, and Woleński [33], [34] pays attention to the ambiguous meaning of the last principle. In the first sense, bivalence means that any sentence is either true or false and thus corresponds to FA_2. In a second sense, bivalence means that any sentence is true or is not. The difference between the two interpretations rests on the meaning of 'false'.[1] A statement can be 'not true' without being 'false' from a many-valued point of view. In response to the many-valued logics promoted by Łukasiewicz, Suszko distinguishes between two kinds of truth-values: *algebraic* values, which are combinations of single truth-values such as 'true-and-false' or 'neither-true-nor-false'; *logical* values, which are sets of values intended to define logical consequence in terms of preserving truth. According to Suszko, there can only be two sets of logical values: *designated* values, which include truth; *non-designated* values, which exclude truth.

From a functional point of view, algebraic values therefore have no interest in providing no essential information to characterize a consequence relation in a given logical system. From an explanatory point of view, on the other hand, we will try to show in this article that the use of algebraic values is likely to shed light on philosophical concepts that a 'Suszkian' logic (without algebraic values) would be unable to explain. The introduction of 'non-Tarskian' or many-valued consequence relations [6], [12] was a first example of this kind, and we will try to see how a constructive approach to truth-values can modify our way to understand some logical and philosophical notions.

There is much more than one way of rejecting FA, if we consider this metalogical axiom as the conjunction of two logically independent propositions. The theory of oppositions can already help us to clarify the situation on this point, by considering FA as a binary proposition of type $FA_1 \wedge FA_2$. With reference to the work of Piaget [13] and Blanché [2], we can affirm that any binary proposition, that is to say, any complex proposition including a binary logical operator, corresponds to a disjunction of four fundamental propositions which are called 'normal conjunctive forms'. Thus, a binary proposition of form $f(p,q)$ refers to four logical possibilities: (i) p and q are true together; (ii) p is true and q is false; (iii) p is false and q is true; (iv) p and q are false together. Frege's position on FA is that both FA_1 and FA_2 are true, while Suszko's position is that FA_1 is false and FA_2 is true. Suszko thus defends a 'counter-thesis', since his position is incompatible with that of Frege. But there is more than one conceivable counter-thesis: that which we will defend in the rest of this work, and which consists in saying that FA_1 is true while FA_2 is false.

Our position is almost the opposite of Suszko's. In fact, there are far more than two FA counter-theses if one considers FA as one of sixteen possible combinations. Indeed, FA means that FA_1 and FA_2 are both true and thus represents a conjunctive proposition such that the two joint members must be true to satisfy the molecular proposition FA. Now fifteen other types of combinations are possible in the light of the theory of propositional conjunctive normal forms. If we use the symbols 1 and 0 to denote the satisfied and dissatisfied normal conjunctive forms, respectively, we obtain the following combinatorial list including the positions of Frege, Suszko, and Łukasiewicz.

	$FA_1 FA_2$	$FA_1 \overline{FA_2}$	$\overline{FA_1} FA_2$	$\overline{FA_1}\, \overline{FA_2}$
(1)	1	1	1	1
(2)	1	1	1	0
(3)	1	1	0	1
(4)	1	0	1	1
(5)	0	1	1	1
(6)	1	1	0	0
(7)	1	0	0	1
(8)	0	0	1	1
(9)	0	1	1	0
(10)	0	1	0	1
(11)	1	0	1	0
(12)	1	0	0	0
(13)	0	1	0	0
(14)	0	0	1	0
(15)	0	0	0	1
(16)	0	0	0	0

Frege's position on FA thus corresponds to (12); that of Suszko corresponds to (14); the position that we will defend, finally, is that represented by (13). As for the thirteen remaining possibilities, their absence from the debate produced between Frege and Suszko is simply due to their non-exclusive form. Formula (2), for example, means an alternative between three possible attitudes: accept FA_1 and accept FA_2, or accept FA_1 and reject FA_2, or reject FA_1 and FA_2. In this sense, this formula symbolizes the union of the three incompatible positions defended by Frege, Suszko, and Łukasiewicz. We can also wonder about the meaning of the two limiting cases (1) and (16): the first consists in admitting all the possible positions about FA_1 and FA_2, while the second consists in admitting none. These kinds of acceptance can be described as 'second order' ones, because they relate to two propositions FA_1, FA_2 whose content is itself accepted or rejected by speakers. Thus Frege accepts (12) because he accepts FA_1 and FA_2, but he rejects the other fifteen formulas by rejecting at least one of the four possible attitudes. Suszko accepts (14) by accepting the attitude of rejecting FA_1 and accepting FA_2, but he rejects all other combinations. We accept (13) by accepting FA_1 and rejecting FA_2, while rejecting all other combinations of attitudes. The distinction between speech orders means that it makes sense to say that a speaker accept to accept, accepts to reject, rejects to accept or rejects to reject any proposition. It also means that there are several levels of discourse, in accordance with the well-established distinction between 'object language' and 'metalanguage'. This semantic abstraction will be important in the rest of the article, especially in relation to the issues of scepticism and the concept of nothingness.

About FA, the presentation of the 16 attitudes in the form of a series of Boolean values 1-0 makes it possible to introduce the theory of oppositions in our present debate. Let us make a first distinction between an antithesis and a counter thesis, which are two logical relations established between propositions or 'theses'. Let $AT(a,b)$ symbolizing the antithesis relation between theses a and b, and let $CT(a,b)$ for the counter thesis relation between a and b. We can then explain these two relations as follows:

- the *antithesis* $AT(a,b)$ means a is contradictory with respect to b and consists in adopting an attitude opposite to it, turning any acceptance of b – symbolized by the value 1 – into a rejection – symbolized by the value 0 (and vice versa);
- the *counter-thesis* $CT(a,b)$ means that thesis a is simply incompatible with the thesis b, turning any attitude of acceptance of b into an attitude of rejection (but the converse need not be the case).

The three positions of Frege, Suszko and Łukasiewicz therefore have only one possible antithesis: AT(12) = (5), AT(13) = (4), and AT(14) = (3). On the other hand, there are as many counter theses to each of these attitudes as there are distinct possibilities of rejecting what is accepted there. In other words, any counter thesis is a thesis which does not accept what the initial thesis accepts but which can reject what the initial thesis rejects. There are thus a total of six counter theses available for the three attitudes of Frege, Suszko and Łukasiewicz:

$$CT(12) = \{(8),(9),(10),(13),(14),(15)\}$$
$$CT(13) = \{(7),(8),(11),(12),(14),(15)\}$$
$$CT(14) = \{(6),(7),(10),(12),(13),(15)\}$$

Since most counter theses are unions of possible attitudes, they have no philosophical relevance to the singular positions (12), (13) and (14). But this first allusion to the theory of oppositions allows at least to sketch a first type of main opposition between three of the sixteen attitudes above: a *triad of contraries* opposing FA (12), the criticism of his first proposition by Suszko (13), and Łukasiewicz's critique of his second proposition (14).

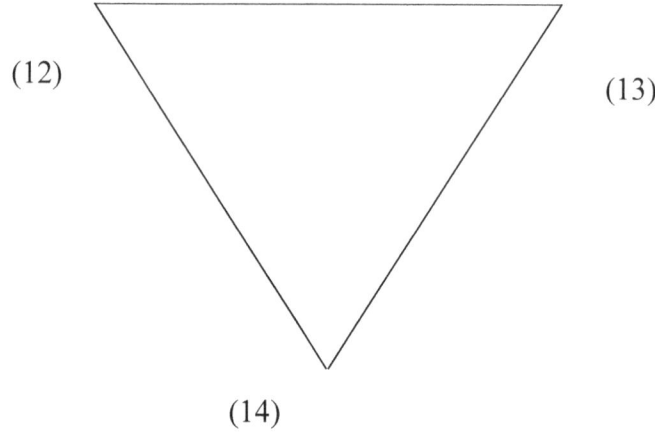

The point is now to examine the content of FA and its two main features, which are functionality (through FA_1) and bivalence (through FA_2).

3. Pragmatic Contradiction

In his attempt to prove one of the fundamental principles of 'classical' logic, i.e. PC, Jan Łukasiewicz [10] has shown not only that it is unprovable but that it also rests on three distinct readings: an 'ontological' reading, by virtue of which PC says that it is impossible for the same object to have a property and not to have it at the same time; a 'logical' reading, whereby PC means that a proposition cannot be true and false at the same time; a 'psychological' reading, by virtue of which PC says that one cannot believe and not believe in the same judgment. The formulation of the logical principle divides those who present it in terms of the truth-value of a proposition and those who formulate it in terms of a proposition and its negation. Woleński [34], [35] will emphasize the two aspects of PC: a logical or 'object language' version, of form

$$\sim(p \wedge \sim p)$$

and a metalogical version, of form

$$v(p) = T \text{ or } v(p) = F$$

Although Łukasiewicz quickly neglected the psychological version of PC, for the reason that subjects often hold inconsistent beliefs,[2] the example given in Łukasiewicz [10] evokes the singular case of religious belief in Trinity: it would be possible according to believe that God is *and* is not the same individual as Father, Son or Holy Spirit, on the occasion of a religious experience that any good Christian would be able to experience within the framework of his faith. Although the empirical objection to PC in its psychological version could be taken seriously, Łukasiewicz was more interested in the ontological and logical foundations of the principle. These do not seem to be more solid than the psychological version, especially since they are based on an ambiguous vocabulary. The ontological principle speaks of objects and refers to facts or states of affairs obtaining in the world. The logical principle speaks sometimes of truth-values sometimes of affirmation and negation, but both cannot make sense without relying on a correspondence theory of truth where facts make a proposition 'true'. As for the 'proposition', it designates from Aristotle onwards any sentence belonging to the grammatical case of indicative and whose linguistic function is to tell something about the world, *viz*. what 'the case' is.[3]

We will not go into the details of this discussion on the foundations of PC, since it goes beyond our central point. We simply observe the following few complications. First, the correspondence theory of truth poses a problem on the conditions of *falsity* of a proposition: either the existence of a fact which contradicts the proposition is necessary to make it false, or the simple absence of fact to make the proposition true entails its falsity. The choice of the correct definition of falsity is important here, since it relates to the universality of PB as well as the validity of PEM. Second, the psychological version of PC makes use of concepts to which everyone else can be reduced. A proposition is true if it corresponds to an objective fact, by virtue of the correspondence theory; but in the absence of sufficient means to prove the existence of such a fact, what is a 'proposition' if not the public expression of a belief expressed by a judgment? On the other hand, there is a common confusion between two pairs of concepts, namely: affirmation and negation, by distinction of truth and falsehood. We know that a proposition can be negative and true, as in 'Poland is not a planet', or affirmative and false, as in 'Poland is a planet'. But what is a judgment, if not the use by a speaker of a proposition in order to sincerely express his own opinion on what 'the case' is? Frege's distinction between a judgeable content and a judgment may be justified, but it seems useless if the correspondence theory of truth is unable to afford the conditions of correspondence with a 'fact' in an incontestable and definitive manner. We know that this theory must face theoretical difficulties, and that two other competing theories of truth face it: truth as coherence, and truth as consensus. Another radical solution could remedy these philosophical difficulties: the 'minimalist' or 'redundant' theory of truth, according to which the occurrence of the concept of truth in a sentence is useless because it does not add any substantial information to it. This last point of view will come back in this article, when questioning the application of Tarski's T-schema.

For want of conclusive answer about the foundations and the validity of PC, let us now try to defend an alternative view and to assess its explanatory virtues: the *pragmatic* (or illocutionary) interpretation of PC, which extends what Łukasiewicz called the 'psychological' version while eliminating its psychological connotation.

We will thus start by assuming that the concepts of truth and falsehood, but also the concepts of affirmation and negation are nothing but items of a general theory of *speech acts*, in which it is not the proposition but the judgment (or statement) which constitutes the primary vehicle of meaning. According to this approach, every statement has the logical form $F(p)$ and includes two elements: a sentential content p, which corresponds to Frege's 'judgeable content'; an illocutionary force F, which carries the purpose that the sentential content is supposed to express in a given dialogue. Since everyday language has the defect of using the same expression for sentential contents, e.g. 'The door is closed', and for their 'assertive' use, let us replace the first with a propositional concept such as 'The door's being closed'. The assertive use of this concept thus yields the speech act 'The door is closed', but there are other uses of the same concept such as the act of questioning, 'Is the door closed?', the act of giving an order, 'Be the door closed!', etc. In the case of the logical principles that concern us here, we can apply this theory of speech acts to lead to some illocutionary interpretations of logical notions.

Affirmation and negation are two types of assertive acts intended to indicate to an interlocutor what 'the case' is, and one of the central points concerns the question of whether these two acts are interdependent or logically independent from each other. 'Truth' and 'falsehood' can be reduced in this theory to the speaker's ontological commitments: to say of a proposition that it is true means that the propositional content it expresses fits to one state of the world; to say that it is false means that it does not fit. There is no difference between the truth of a proposition and the recognition of its truth by the speaker, within the framework of this theory. If this is the case, the problem is to know if this speaker can act other than by recognizing the truth of what a statement expresses. There can be but one judgment, according to Frege: either we recognize the truth of a statement, and we express the latter by 'The door is closed'; either we do not recognize it, and we express its falsity indirectly by 'The door is not closed'. But what if the speaker does not know whether the door is closed?

Von Wright proposes a grammatical test to know if a statement is a 'proposition': "A grammatically well-formed sentence expresses a proposition if, and only if, the sentence which we get by prefixing to it the phrase 'it is true that' is also well-formed" [29, p. 6.][3] For example, 'It is true that the door is closed' is well-formed and, therefore, the sentence 'The door is closed' is a proposition; on the other hand, that 'It is true that close the door!' is an ill-formed sentence entails that 'Close the door!' is not a proposition. Von Wright's analysis does not just corroborate the theory of speech acts anticipated by Aristotle and established by Searle; it will also justify the existence of propositions which are neither true nor false, such as normative propositions like 'The length of the standard meter in Paris is 1 meter'[4] or metaphysical propositions like 'To be is to be perceived'. A proposition can therefore be neither true nor false while belonging to the class of assertive acts. We will examine in the following the consequences of this result on several issues of logic, all scrutinized by Woleński.

4. Pragmatic Bivalence

Woleński [34] presents PB as the conjunction of two metalogical propositions. The first is a metalogical version of PEM:

(1) Every sentence is either true or false.

The second is a metalogical version of PC:

(2) No sentence is both true and false.

The combination of (1) and (2) relates to four forms of metalogical judgments, either affirmative or negative. By taking up the idea of von

Wright [29], i.e. prefixing arbitrary sentences p by a truth operator T, we thus obtain four types of judgments: Tp for 'It is true that p' and Fp for 'It is false that p', which are affirmative judgments; ~Tp for 'It is not true that p' and ~Fp for 'It is not false that p', which are negative judgments. As usual, Woleński [34] then proposes a logical hexagon, (S2), to represent the logical relationships between these four judgments. The advantage of this hexagon is that it relativizes PB by depicting it as a non-tautological vertex (see below).

The top vertex of (S2) expresses the affirmative clause (1) of PB, while its contradictory at the bottom vertex symbolizes the negation of (1). Consequently, anyone who subscribes to PB cannot think in terms of this hexagon without accepting two situations incompatible with bivalence: on the one hand, the possibility for a sentence not to be true without being false (~Tp does not entail Fp) and not to be false without being true (~Fp does not entail Tp); on the other hand, there is the possibility for a sentence to be neither true nor false (~Tp & ~Fp). At the same time, (S2) does not include the case of 'true contradictory' sentences, or *dialetheias*, of form T$p \wedge$ Fp. This sentence is incompatible with (S2), since Tp and Fp are contrary to it (~(T$p \wedge$ Fp) holds in (S2)). However, just as what Woleński calls 'neutralities' [34, p. 103] is possible in (S2) and consists in rejecting the affirmative clause of PB, an opponent of PB must be able to accept the possibility of 'dialetheias' and to reject (2) into a non-bivalent logic. Although Woleński doubts the intuitive meaning of dialetheias by claiming that "I did not find any natural matrix semantics for paraconsistent logic" [34, p. 12], it is nevertheless possible to justify their existence, in particular by proceeding with what von Wright [29] describes as a shift of meaning in the concept of truth.

Referring to the example of drizzle as a 'transition zone' between rainy and dry weather, von Wright explains that this situation can be logically analyzed in two distinct ways: either as a situation where it is neither totally the case that it rains nor totally the case that it does not rain, insofar as drops of water still fall from the sky; either as a situation where it is still raining and already the case it is not raining anymore, insofar as simple drops of water are still rain and already a situation of no rain. This means that one and the same situation can be considered either as a case of neutrality or as a case of dialetheia, but not by virtue of the same interpretation of what 'the case' is or truth:

> It should be observed that a conceptual shift has now taken place in the notion of truth. It is not the same sense of 'true' in which we say that is neither raining nor not-raining and say that it is both raining and not-raining in the zone of transition. We could call the former a *strict* sense of 'true' and the latter a liberal or *more lax* sense of truth [29, p. 13].[5]

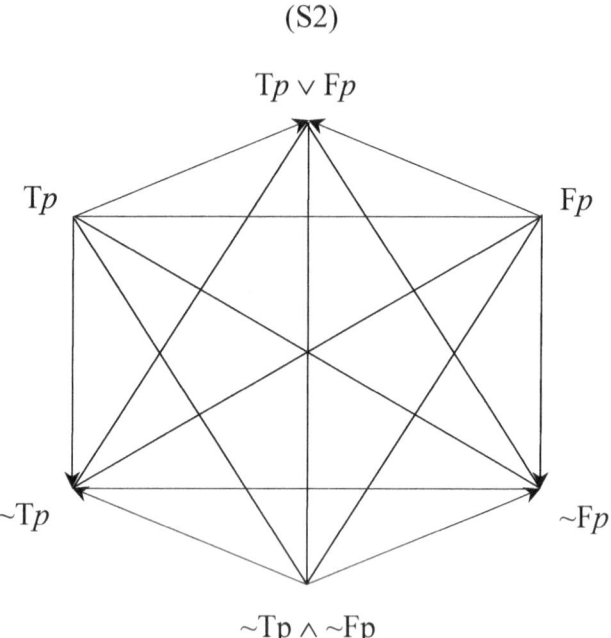

The difference between the two interpretations of the concept of truth appears clearly in the below diagram by Von Wright [29], [30], where the 'gappy' and 'glutty' propositions do not correspond to the same 'logical zone' and are expressed respectively by a strict operator T and a liberal operator T'.

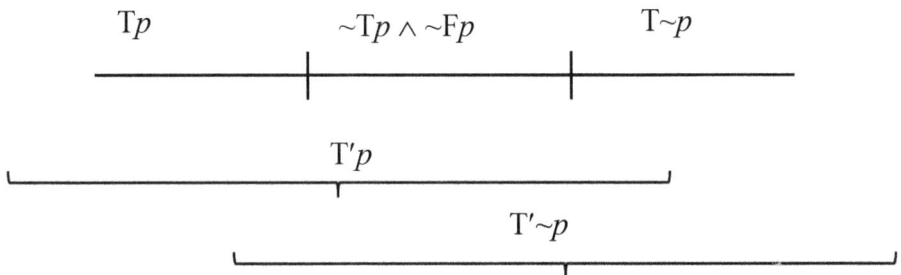

Woleński's hexagon (S2) is therefore only a fragment of non-bivalent logical systems, and we must take into account the two interpretations of the concept of truth in order to establish a set of exhaustive relationships between all possible judgments of truth and falsehood.

For this purpose, we propose in the following a reconstruction of PB in the form of four independent clauses. Indeed, (1) and (2) above are complex formulas including two atomic sentences. (1) can be divided into two conditional sub-sentences or clauses implying an affirmative consequent:

(1.1) If p is not false, then p is true.
(1.2) If p is not true, then p is false.

Similarly, (2) can be divided into two clauses implying a negative condition:

(2.1) If p is true, then p is not false.
(2.2) If p is false, then p is not true.

These four clauses are of a metalogical order, so they are no material conditionals; rather, they represent procedural constraints imposed on affirmative judgments, such that it is not allowed to assign a certain algebraic value to a sentence without rejecting another at the same time. The general form of PB is a clause like

$$\text{If } p \text{ is } X, \text{ then } p \text{ is not } Y$$

and can be interpreted as a *mapping*, that is, a homomorphism between a domain of values and a counter-domain:

$$X \mapsto Y$$

where X and Y denote arbitrary truth-values within a given domain.

An important question is: can a rational agent subscribe to one or the other of these clauses without admitting the others, so that each of them would be considered logically independent of all the others? A reconstruction of PB was recently proposed in Schang [25], where this last principle is represented as the combination of four statement operators $[A_i]p$, meaning 'It is the case that p' or 'It is true that p'. Assertion corresponds to the assertive speech act by which the agent accepts the truth of p, in accordance with our pragmatic interpretation of the concept of truth. This operator is also able to translate the two distinct meanings of the concept of truth expressed by von Wright, while preserving the idea accepted since Frege according to which affirming the falsehood of a sentence is affirming the truth of its sentential negation,

$$Fp \Leftrightarrow T{\sim}p$$

In addition, we will see that these two 'truth-operators' are only two particular cases within a general logic of acts of acceptance and rejection.[6] On the basis of the *partial* statement operator $[A_i]$, PB can be reconstructed as a set of four types of operators applied to truth-values and translating the four clauses (1.1) – (2.2) as follows:

(1.3) $[A_1]p: \overline{T} \mapsto \overline{F}$
(1.4) $[A_2]p: \overline{F} \mapsto \overline{T}$
(1.1) $[A_3]p: \overline{T} \mapsto F$
(1.2) $[A_4]p: \overline{F} \mapsto T$

These operators are 'partial', because they transform only some of (but not all) values of the initial domain: if p is X, then p is not Y; but if p is not X in the initial state of the domain of values, then nothing happens, i.e. no transformation occurs in the final state of the counter-domain.[7] The 'positive' values T and F denote acts of acceptance (of truth and falsehood), as opposed to the 'negative' values which denote acts of rejection (of truth or falsehood). The independence of negative values is explained by our pragmatic interpretation of truth-values and by the primacy of acts of judgment over the assignments of these truth-values, which are only expressions of propositional attitudes towards a primary truth-value: the True. Von Wright explains this point as follows:

> How many truth-values are there? Shall we say there are two: truth, and falsehood? Or count the gaps and overlaps too as truth-values and say there are four in all? As will be seen later, we shall make use of a 4-valued matrix. But since all four values are definable in terms of *truth* and *negation*, it would also be possible to say that basically there is only *one* 'truth-value', viz. true [29, p. 314].[8]

Moreover, 'classical' logic seems 'simpler' than four-valued systems insofar as it makes uses of only two truth-values. But from our pragmatic point of view, it is less simple because it imposes more constraints on the acts of acceptance and rejection. The Bivalentist equates the 'false' with any statement that is not true, so that rejecting the truth of a statement is sufficient to accept its falsity from his point of view. However, this condition is not imposed on a speaker whose rationality does not include the operator [A$_3$].

As a matter of fact, there is a set of 2^{m-1} possible acceptance operators within a domain of m truth-values, knowing that this domain of values can increase from $2^n = m$ to 2^{n+1} elements.[9] Woleński [34] pointed out that any domain of values including $m < 2$ truth-values is trivial and unable to satisfy the properties of the Tarskian consequence operator. In the present case, let us consider the particular domain of values in which the $m = 4$ truth-values are the true T, the non-true $\overline{T}$, the false F, and the non-false $\overline{F}$. The $2^4 - 1 = 15$ acceptance operators available in this domain are the following, where the product $\oplus$ consists in adding a variable number of restrictions on judgments.

$$[A_1]p: T \mapsto \overline{F}$$
$$[A_2]p: F \mapsto \overline{T}$$
$$[A_3]p: \overline{T} \mapsto F$$
$$[A_4]p: \overline{F} \mapsto T$$

$$[A_5]p = ([A_1] \oplus [A_2])p: T \mapsto \overline{F} \oplus F \mapsto \overline{T}$$
$$[A_6]p = ([A_1] \oplus [A_3])p: T \mapsto \overline{F} \oplus \overline{T} \mapsto F$$
$$[A_7]p = ([A_1] \oplus [A_4])p: T \mapsto \overline{F} \oplus \overline{F} \mapsto T$$
$$[A_8]p = ([A_2] \oplus [A_3])p: F \mapsto \overline{T} \oplus \overline{T} \mapsto F$$
$$[A_9]p = ([A_2] \oplus [A_4])p: F \mapsto \overline{T} \oplus \overline{F} \mapsto T$$
$$[A_{10}]p = ([A_3] \oplus [A_4])p: \overline{T} \mapsto F \oplus \overline{F} \mapsto T$$

$$[A_{11}]p = ([A_1] \oplus [A_2] \oplus [A_3])p: T \mapsto \overline{F} \oplus F \mapsto \overline{T} \oplus \overline{T} \mapsto F$$
$$[A_{12}]p = ([A_1] \oplus [A_2] \oplus [A_4])p: T \mapsto \overline{F} \oplus F \mapsto \overline{T} \oplus \overline{F} \mapsto T$$
$$[A_{14}]p = ([A_2] \oplus [A_3] \oplus [A_4])p: F \mapsto \overline{T} \oplus \overline{T} \mapsto F \oplus \overline{F} \mapsto T$$

$$[A_{15}]p = ([A_1] \oplus [A_2] \oplus [A_3])p: T \mapsto \overline{F} \oplus F \mapsto \overline{T} \oplus \overline{T} \mapsto F \oplus \overline{F} \mapsto T$$

Von Wright's *strong* truth-operator T corresponds to the acceptance operator $[A_6]$: it consists in judging as true every sentence that is not held false, and as false every sentence that is not held true.

$$Tp = [A_6]p: T \mapsto \overline{F} \oplus \overline{T} \mapsto F$$

The operator T thus obeys two of the four clauses of PB, *i.e.* (1.1) and (2.1). The *weak* truth-operator T' corresponds to another acceptance operator, $[A_9]$:

$$T'p = [A_9]p: F \mapsto \overline{T} \oplus \overline{F} \mapsto T$$

according to which any sentence that is held false is a sentence that is not held true and any sentence that is not held false is a sentence that is held true. Hence the operator T' does not obey the same clauses of PB as the operator T, since the latter satisfies (1.2) and (2.2).

By interpreting pragmatically the four truth-values as an ordered pair of accepted or rejected truth-values, we thus obtain the translations $B = 11$, $T = 10$, $F = 01$ and $N = 00$ and the following matrices characterizing the 'strict' and 'liberal' truth-operators.

p	Tp	$T'p$
11	10	01
10	10	10
01	01	01
00	01	10

The third metalogical operator of von Wright [29,30], the operator of falsity Fp, does not occur among the operators $[A_1] - [A_{15}]$ because its definition is essentially based on a type of information irreducible to the terms of PB. It essentially involves sentential *negation*, knowing that $Fp \Leftrightarrow T{\sim}p$. It is this negation that we will explain now, in pragmatic terms of *rejection*.

The second type of judgment, *viz.* rejection [N], is independent of the acceptance operator $[A_i]$. The latter imposes restrictions on 'positive' judgments of type 'p is X', while rejection corresponds to the class of operators imposing constraints on 'negative' judgments of type 'p is not X'. The general form of the rejection operator is

if p is X, then p is not X,

that is, a mapping of the general form

$$X \mapsto \overline{X}$$

which differs from the operator of acceptance by the identity of the transformed truth-value of the counter-domain. Due to the procedural similarity in the mappings of affirmative and negative judgments, there may be as many separate rejection operators [N_i] as there are acceptance operators [A_i]. To build such a rejection operator, it suffices to repeat the pattern of [A_i] whilst replacing the truth-value of the counter-domain Y by the value X of the initial domain.

'Classical' negation can be understood in two distinct ways, in this general logic of acceptance and rejection: either as that which turns the true into false and the false into true, by virtue of PB; either as that which turns the true into non-true and the false into non-false, independently of PB. Suszko's acceptance of the clause FA_2 consists in treating the two explanations above as equivalent: the algebraic 'not-true' is a logical 'false', in the sense that it expresses a *not-designated* value $\bar{D}$ which excludes the algebraic 'true'; the algebraic 'not-false' is a logical 'true', in the sense that it expresses a *designated* value D which includes the algebraic 'true'. Now there are rejection operators [N_i] which do not turn any designated value into an undesignated value, and vice versa. Taking the example of the particular operator [N_2],

$$[N_2]p: F \mapsto \bar{F}$$

this case of partial rejection is such that the initial truth of p is left unchanged in the final counter-domain. It is therefore necessary to specify the formulation of 'classical' negation [N_C] as any rejection operator which turns the designated or 'true' into non-designated or 'non-true', i,e.,

$$\{[N_C]p \mid [N_C]p \supseteq T \mapsto \bar{T}\}$$

However, 'classical' negation behaves traditionally as a 'total' function that transforms *all* the truth-values of the initial domain; thus, its pragmatic counterpart will be here the rejection operator [N_{15}]:

$$[N_{15}]p = ([A_1] \oplus [A_2] \oplus [A_3] \oplus [A_4])p: T \mapsto \bar{T} \oplus F \mapsto \bar{F} \oplus \bar{T} \mapsto T \oplus \bar{F} \mapsto F$$

We will call this total negation a 'Boolean' negation, rather than a 'classic' negation which designates the use of negation within the classical logic system.

Falsehood, on the other hand, is a 'mixed' operator that associates the operator of truth with a sentential negation. The negation in question in the operator of falsehood Fp = T$\sim p$ is what von Wright [29] describes as *internal* or strong negation ('it is the case that *not*'), as opposed to *external* or weak negation or weak ('it is *not* the case that'). Internal negation is not prefixed to the operator T, but to the sentential content p. Now the rejection operators [N$_i$] are not able to explain this negation, because they are only constructors of external negations. Strong negation stands 'halfway' between the operators of acceptance and rejection, insofar as it consists in accepting the negation of a given statement and not in simply denying this statement. To represent strong negation, we need a third type of mapping which is neither acceptance nor rejection but a 'fusion' of the two basic judgment operators. This hybrid operator can characterize internal negation (or 'Morganian') as follows:

$$[AN_i]p: X \mapsto \overline{\overline{Y}} = X \mapsto Y$$

which can be paraphrased as 'rejected acceptance' or 'accepted rejection'[10] and whose traditional characterization corresponds to total negation [AN$_{15}$]:

$$[AN_{15}]p: T \mapsto F \oplus F \mapsto T \oplus \overline{T} \mapsto \overline{F} \oplus \overline{F} \mapsto \overline{T}$$

It is therefore possible to translate the operator of falsity F = T$\sim p$ as the expression of a particular affirmation

$$Fp = T[NA_{15}]p = [A_9][NA_{15}]\,p,$$

and weak truth T'p = $\sim$Fp as the total rejection of the falsity-operator:

$$T'p = \sim Fp = [N_{15}][A_9][NA_{15}]p.$$

We obtain the characteristic matrices of the operators F and T' on the basis of their above pragmatic reconstruction:

p	$[NA_{15}]p$	$[A_9][NA_{15}]p$	$[N_{15}][A_9][NA_{15}]p$
11	11	10	01
10	01	01	10
01	10	10	01
00	00	01	10

The set of operators of acceptance, rejection, and the fusion of both constitutes a generalized logical framework, $AR_{4[Oi]}$,[11] which is a set of 4-valued systems composed of the usual logical constants (conjunction, disjunction, conditional), pragmatic operators $[O_i] = \{[A_i],[N_i]\}$, and in which the sentential variables (atomic or molecular) are always preceded by a judgment operator $[O_i]$.[12] An extension of this framework to n-valent systems corresponds to the universal logical framework $AR_{4[Oi]}$, but our reflection will be limited to the 4-valued domain in the following.

It is already possible to reconstruct four types of logical systems in $AR_{4[Oi]}$, by means of the two operators of acceptance and rejection and the two negations, external (or 'Boolean') and internal (or 'Morganian'). These three categories are distinguished by their attitude towards two main metalogical properties: *completeness* (semantics), and *consistency*. The first property corresponds to the clauses (2.1) – (2.2) of PB, and the second property corresponds to the clauses (1.1) – (1.2).

The first category of logical systems translatable in terms of $[A_i]$ is the set of 'classical' systems, that is to say, complete and consistent. They correspond to von Wright's logic CL. Although there is traditionally only one single logic system called 'classical', there can exist more than one if we interpret the term 'classical' in the sense of Suszki's bivaluation: every sentence receives only one truth-value, T or F, in accordance with the second clause FA_2 of Frege's Axiom. The class of 'classical' systems thus corresponds to the class of logical systems whose characteristic operators $[A_i]$ produce only two algebraic values: those which von Wright calls 'unilateral' truth, 10 (true and non-false) and 'unilateral' falsehood, 01 (false and not-true). It has been shown in Schang [25] that several $[A_i]$ are consistency-and-completeness-forming operators in that they form 'unilateral' judgments: $[A_6]$, $[A_7]$, $[A_8]$ and $[A_9]$, whose common feature is to satisfy one and only one clause of consistency and completeness: (1.1) – (2.1), (1.1) – (2.2), (1.2) – (2.1), or (1.2) – (2.2). The matrices below show the 'classic' behavior of rational agents subscribing to one or other of the bivalent restrictions on truth-values:

p	$[A_6]p$	$[A_7]p$	$[A_8]p$	$[A_9]p$
11	10	10	01	01
10	10	10	10	10
01	01	01	01	01
00	01	10	01	10

The second category of logical systems is the set of complete and non-consistent systems, *viz.* 'paraconsistent'. They correspond to von Wright's logic T'L. Each of these systems satisfies one completeness clause among (1) and (2), but none of the two consistency clauses (3) – (4). There are again several ways of obtaining these conditions in $AR_{4[Oi]}$, hence several paraconsistency-forming operators of acceptance: $[A_3]$, $[A_4]$, $[A_{10}]$, $[A_{13}]$, $[A_{14}]$, whose common feature is to accept the 'glutty' or 'overlapping' algebraic value $B = 11$.

p	$[A_3]p$	$[A_4]p$	$[A_{10}]p$	$[A_{13}]p$	$[A_{14}]p$
11	11	11	11	11	01
10	10	10	10	10	10
01	01	01	01	01	01
00	01	10	11	11	11

The third category of logical systems is the class of non-complete and consistent systems, or 'paracomplete'. They correspond to von Wright' TL logic. Each of these systems satisfies one consistency clause among (2.1) and (2.2), but none of the two completeness clauses (1.1) – (1.2). The paracompleteness-forming operators of acceptance admit the 'bilateral' incomplete value $N = 00$ and are the following: $[A_1]$, $[A_2]$, $[A_5]$, $[A_{11}]$, $[A_{12}]$.

p	$[A_1]p$	$[A_2]p$	$[A_5]p$	$[A_{11}]p$	$[A_{12}]p$
11	10	01	00	00	00
10	10	10	10	10	10
01	01	01	01	01	01
00	00	00	00	01	10

Finally, the fourth and final category of logical systems is the class of non-complete and non-consistent, or 'paranormal' systems. They correspond to von Wright's logic T"L. Paradoxically, these systems are not those which satisfy none but, on the contrary, *all* the four clauses (1) – (2) of PB. The paranormality-forming operators of acceptance admit the two 'bilateral' values $B = 11$ and $N = 00$ and are reduced to one single case: $[A_{15}]$.

$[p]$	$[A_{15}]p$
11	00
10	10
01	01
00	11

The above results partially agree with the classification proposed by Woleński for the different attitudes towards PB and Suszko's Thesis, which consists in dividing any language L into two and only two classes of logical values. While we recognize that there are different ways of disagreeing with PB, we are not following the same classification criteria. According to Woleński, the disagreement relates to 'Bivalentists', the 'Pseudobivalentists', and the 'Antibivalentists':

> The Bivalentists accept PB (the conjunction of (1) and (2)), but they differ as far the matter concerns whether the bi-division of L suffices for constructing logic. The Pseudobivalentists accept either the metalogical *tertium non datur* (1) or the metalogical principle of non-contradiction (2) and take the bi-division as sufficient or not. The Antibivalentists accept neutralities or dialetheias and deny that the bi-division adequately displays the basis of logic [34, p. 105.].

From a constructive point of view, Woleński's Bivalentists 'are these 'Semi-bivalentists' who form classical, that is to say, complete and consistent judgments; the Pseudo-bivalentists are these 'Semibivalentists' who form paracomplete or paraconsistent judgments; and the Antibivalentists are, paradoxically, the 'complete' Bivalentists who form paranormal judgments by admitting the four clauses of PB. This paradoxical result comes from our 'constructive' or analytical reading of bivalence, while Woleński [34] does not divide the clauses of consistency and completeness into two logically independent clauses. In all cases, the bi-division required by Suszko's Thesis never allows the construction of non-classical logic systems in the framework of $AR_{4[Oi]}$.

Our pragmatic reconstruction of logical systems hopes to draw attention to four main points.

Firstly, 'classical' logic and 'bivalent logic' are not synonymous expressions from our pragmatic point of view. The so-called 'classical' logic was constructed by 'semi-bivalent' systems, insofar as it does not satisfy the four PB clauses but only two of them (as opposed to the 'paranormal' logic which, paradoxically, obeys all clauses of PB but is not a 'classical' system).

Secondly, there are strictly speaking no 'classical', 'paracomplete' or 'paraconsistent' negations. It is shown above and in Schang [25] that it is not the two sentential negations (Boolean and Morganian) but the acceptance operators that distinguish the classes of theorems from classical and non-classical systems. In other words, the 'classical' agent is not distinguished from other agents by his particular use of negation but, rather, by his attitude towards PB or what justifies a statement of truth.

Thirdly, the illocutionary interpretation of judgments provides a certain answer to the sea-battle problem, presented in the Chapter IX of Aristotle's *De Interpretatione* and studied a length by Łukasiewicz. To the question of how to validate PEM,

PEM Every sentence p or its negation $\sim p$ is true,

without admitting the completeness clause of PB,

(2) Every sentence is true or false,

Łukasiewicz [11] proposed a trivalent logic which rejects PB but is not able to validate PEM. Is it only possible, and on what condition in $AR_{4[Oi]}$? The problem essentially relates to the relation between PEM and its metalogical version of *tertium non datur*, expressed by the completeness clause (2) of PB. If the two principles are independent from each other, it means that it is possible to admit one without the other. In accordance with the formation rules of judgments and the translation of PEM and (2) in $AR_{4[Oi]}$:

(PEM') $\qquad [A_i](p \vee [N_{15}]p)$

(2') $\qquad [A_i]p \vee [A_i][N_{15}]p$

validating PEM without BV consists in finding an interpretation of $[A_i]$ such that the following thesis is not valid:

$$(PEM') \rightarrow (2')$$

The operator $[A_8]$ seems to satisfy this request but requires an extension of Łukasiewicz's analysis to four-valuedness. By replacing sentential variables with their algebraic referents, we thus obtain the following proof of invalidity:

$$[A_8]11 \vee [N_{15}]11 \rightarrow [A_8]11 \vee [A_8]00$$
$$[A_8](11 \vee 00) \rightarrow [A_8]11 \vee [A_8]00$$
$$[A_8]10 \rightarrow 01 \vee 01$$
$$10 \rightarrow 01$$
$$01$$

One can doubt, however, the philosophical insightfutlness of this result, beyond its purely formal meaning. Even if the paranormal situation of the antecedent may agree with the thesis of indeterminism, the problem, on the other hand, concerns the meaning of sentential negation: the proof above rests essentially on the use of *Boolean* negation, while the notion of falsehood included in the consequent PB results in a *Morganian* negation which modifies the above result of the proof. Another solution would be to admit bivalence by replacing the notion of falsehood with that of non-truth, so that the consequent PB would be replaced by PB'. Now such a formal 'solution' does not account for the sea-battle problem by reducing it to a trivial version. It therefore remains an open issue, especially regarding the meaning to be given to sentential negation: Morganian, in 'it is the case that the sea-battle will take place or will *not* take place'; or Boolean, in 'it is the case that the naval battle will take place or it is *not* the case that the sea-battle will take place'.[13]

Fourthly, the functional definition of the acceptance operators provides a new and completely abstract explanation of the concept of *duality*. Thus, for any operator $[A_i]$ and all algebraic values X,Y:

$$\text{If } [A_i]p: X \mapsto Y, \text{ then } d([A_i]p): \overline{X} \mapsto \overline{Y}$$

Each of the paracomplete systems of $AR_{4[Oi]}$ is the dual of a paraconsistent system, by this definition. Duality can also be interpreted by the relation of *subalternation*, within the framework of the theory of oppositions. It is this notion of duality that we will find again later on, with respect to the relationship between logic and scepticism. It will allow to see if there are other types of propositional attitudes than acceptance and rejection or if attitudes such as assertion, assumption and doubt are all reducible to the two pragmatic operators of $AR_{4[Oi]}$.

5. Logics of Attitudes

If we accept the hypothesis that logic concerns the relationship between judgments, this implies that the speaker commits to the truth of certain sentences and also accepts the logical consequences of this commitment. However, the preceding discussion on bivalence and the plurality of agents represented in $AR_{4[Oi]}$ assumes that these agents are multiple and that not only one rationality is assumed by all of these. Woleński draws a relevant formal lesson from this plurality, with respect to the redundant theory of truth and Tarski's T-scheme: "considerations about the T-scheme show that T-equivalences are no longer logical tautologies beyond propositional calculus" [35, p. 9.][14]

Von Wright confirms that the T-scheme does not hold universally

> The equivalence $Tp \leftrightarrow p$ is well-known from discussions about the nature of truth. Its meaning is often expressed by saying that the phrase 'it is true that', when prefixed to a sentence, is otious or redundant. But this is true only if one accepts the laws (of excluded middle and of contradiction) of classical logic. In classical logic the phrase 'it is true that' is indeed redundant – and this explains why the truth operator is not needed in the object language of the classical calculus. But the classical calculus is only a special, limiting, case of truth-logic. In other truth-logics the truth-operator is not redundant [30, p. 325].[15]

This is the obviously the case in the pluralist framework $AR_{4[Oi]}$, where the T-scheme may not be valid for some interpretations of $[A_i]$.[19]

Faced with a plurality of formal truth-logics, Woleński [30], [32] investigates the philosophical forms of this plurality. He goes on discussing its traditional expressions in the history of philosophy, especially through the distinction established by Sextus Empiricus between three patterns of rationality: dogmatism, academicism, and scepticism. The latter seems to pose an enigma for logic: Does the sceptical agent recognize some particular cases of truth, and what logic can he admit if he does not recognize any? Let us take a look at Woleński's analysis, in order to see what we can learn from it within our pragmatic logic.

This analysis is based on three components: a precise definition of the three philosophical schools mentioned above, based on the commentary by Arne Naess; a representation of the logical relationships between the three types of agents which result from it, in the form of a hexagon of logical oppositions; an explanation of sceptical logic 'in the form of a dual consequence'.

The main trouble comes from the meaning of the original text by Sextus Empiricus. According to the author, each of the three philosophical schools studied expresses a distinct *epistemic attitude:*

dogmatism asserts that the truth is discoverable; academicism asserts that the truth is not discoverable; the sceptic does not assert that the truth is discoverable, nor does he assert that the truth is not discoverable (he seeks the truth without finding it yet). We recognize here another triad of contraries between the three positions, and Woleński offers an exhaustive representation of epistemic attitudes in the form of a logical hexagon (DIA),[16] isomorphic to the preceding hexagon of truth-values (S2).

The inclusive position (ε) corresponds to general dogmatism and comes in two forms: (α) 'optimistic' dogmatism, expressed by academicism; (β) 'pessimistic' dogmatism, expressed by academicism. The exclusive position (φ) corresponds to scepticism, which rejects the positive assertion of the dogmatist and the negative assertion of the academician.

(DIA)

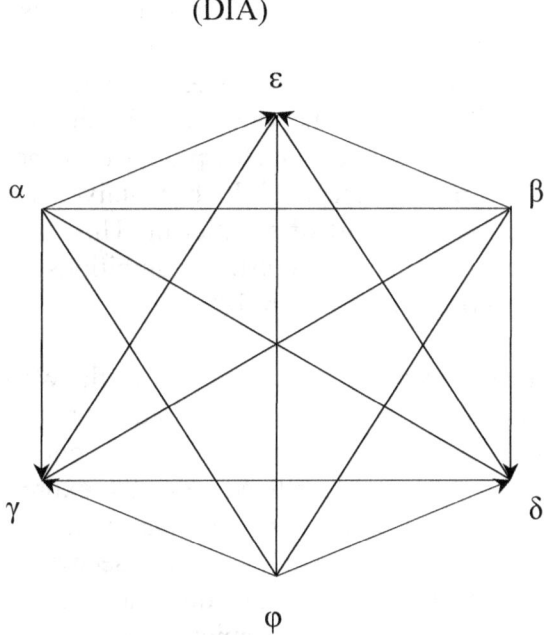

The first logical difficulty comes from the meaning to be given to the indefinite expression '*the* truth': is it any truth whatsoever, or some truth in particular? According to Woleński, only an existential interpretation of this definite article can restore the precise meaning of the academic position:

> The dogmatist's view cannot be rendered by
> (20) I assert that *every* truth is discoverable,
> because it would make it impossible to state academism and scepticism adequately to their actual historical form. Assume that (20) is taken as proper for dogmatism. By our (DIA), the academician would say
> (21) I assert that *not every* truth is discoverable (= I assert that at least one truth is not discoverable). However, this statement is too weak for the academician, because it does not exclude that possibly some truths are discoverable. Now the sceptic, under (20) and (21), must say
> (22) I do not assert that every truth is discoverable and I do not assert that at least one truth is not discoverable.
> This statement is too weak for the sceptic, because it ascribes to him the view that abstaining from assertions is restricted only to selected propositions belonging to a given K. Since the sceptical doubt is universal in (K), (22) drops an essential part of scepticism. This, the dogmatist should be moderate in his epistemic ambitions in order to be fair to his competitors [33, p. 189.]

This passage seems to express a half-truth. Woleński is right to say that the sceptic does not accept any particular truth according to the principle of isosthenia, so that there is no sufficient reason to assert p or its negation $\sim p$. On the other hand, does the expression of sceptical attitude impose one and only one possible expression for the attitudes of dogmatism and scepticism? Woleński seems to think so. His reasoning proceeds as follows: if the dogmatist asserts that *every* truth is discoverable, then the academician asserts the contrary. What is this contrary? According to Naess, this is the assertion that at least one truth is not discoverable; however, this assertion does not correctly restore the attitude of the academician; therefore, the dogmatist's attitude must be reformulated accordingly and expressed as the assertion that *at least* one truth is discoverable. This reasoning is based on the idea that the triad of opposites αβφ of the hexagon (DIA) must be exhaustive ; that is to say, it must exhaust the entire space logic so that $\alpha \vee \beta \vee \varphi$ is a *tautology*.

At the same time, it is possible to express a greater number of epistemic attitudes than those expressed in (DIA). As shown by Englebretsen [5], this number depends on the logical structure of the expressions and the different ways of denying them. Starting from

(a)   I assert that every truth is discoverable,

it is possible to express seven other different judgments on the basis of (a), modifying its logical form by the introduction of negations:

(b)   I assert that every truth is *not* discoverable.
(c)   I assert that *not* every truth is discoverable, i.e. I assert that at least one truth is not discoverable.
(d)   I do *not* assert that every truth is discoverable.
(e)   I assert that *not* every truth is *not* discoverable, i.e. I assert that at least one truth is discoverable.
(f)   I do *not* assert that *not* every truth is discoverable.
(g)   I do *not* assert that every truth is *not* discoverable.
(h)   I do *not* assert that *not* every truth is *not* discoverable.

The logical space of the formulas (a) – (h) is more complex than the hexagon (DIA), due to the logical structure of its formulas. In (DIA), negation is restricted and applies only to the sentential content 'every truth is discoverable'. The assertive *modality* of judgment is never denied, while it is in (d), (f), (g), and (h). By analogy with the alethic modalities of necessity and possibility, assertion can be considered as a 'strong' epistemic modality and its negation means the 'weak' modality of *supposition*. Scepticism denies positive and negative assertions, therefore the attitude of doubt that characterizes it is equivalent to an epistemic contingency. On the basis of this interpretation, we can reformulate the negative judgments as follows:

(d) I suppose that *not* every truth is discoverable, i.e. I suppose that at least one truth is *not* discoverable;
(f) I suppose that every truth is discoverable;
(g) I suppose that *not* every truth is *not* discoverable, i.e. I suppose that at least one truth is discoverable;
(h) I suppose that every truth is *not* discoverable.

The diagram (DIA) is therefore a mere fragment of this set of expressions in which (α) = (e), (β) = (b), and (φ) = ~(e) ∧ ~(b) = (h) ∧ (g).

The problem to follow is twofold: What are the logical relationships between the formulas of the extended structure (a) – (h)? Did Naess and Woleński provide a correct interpretation of the three epistemic attitudes of dogmatism, academicism, and scepticism?

To study the set of logical relationships, we may analyze the logical space of these formulas as a set of logically independent subsets, that is to say mutually exclusive and exhaustive. The result is a *Partition Semantics*, similar to the analysis, of FA proposed in section 1 and inspired by various works [4], [9], [24]. The hexagon (DIA) is limited to a logical space Σ_1 composed of three subspaces:

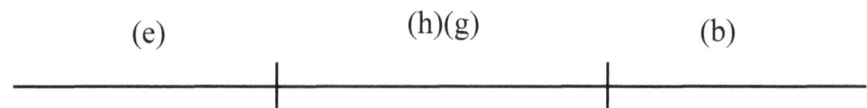

while the logical space Σ_2 of the expressions (a) – (h) includes six subspaces which are further partitions of the three previous ones:

Each expression can then be interpreted as a set of occupied or unoccupied positions into a finite logical space. Let σ be the function applying to each expression a corresponding value 1 or 0 in the different logical subspaces. This results in the following valuations for all of the formulas (a) – (h), to be identified by a characteristic *bitstring* (an ordered sequence of Boolean bits):

σ(a) = 100000, σ(b) = 000001, σ(c) = 001011, σ(d) = 011111, σ(e) = 111000, σ(f) = 110100, σ(g) = 111110, σ(h) = 110111.

The above valuations above make it possible to define the set of logical relations by means of a Boolean calculus, composed of three bitstring operators: complementation, union, and intersection. Thus, for any bitstring σ(x) = ⟨σ₁(x), ..., σₙ(x)⟩ of length n characterizing any abstract object x:[17]

Complementation
$\overline{\sigma(x)} = \langle \overline{\sigma_1(x)}, \ldots, \overline{\sigma_n(x)} \rangle$

Union
σ(x) ∪ σ(y) = ⟨σ₁(x) ∪ σ₁(y), ..., σₙ(x) ∪ σₙ(y)⟩, with 1 > 0 and $\sigma_i(x) \cup \sigma_i(y) = \max(\sigma_i(x), \sigma_i(y))$.

Intersection
σ(x) ∩ σ(y) = ⟨σ₁(x) ∩ σ₁(y), ..., σₙ(x) ∩ σₙ(y)⟩, with 1 > 0 and $\sigma_i(x) \cap \sigma_i(y) = \min(\sigma_i(x), \sigma_i(y))$.

Following the calculus of oppositions presented by Schang [24], complementation turns out to be a *contradiction*-forming operator. If the definitions of Naess and Woleński are correct, then:

- ('positive') dogmatism is (e), and its characteristic bitstring in Σ_2 is σ(e) = 111000
- academism (or 'positive dogmatism') is (b), and its characteristic bitstring in Σ_2 is σ(b) = 000001
- scepticism is a negation of the two dogmatisms, therefore it corresponds to the formula
  ~(e) ∧ ~(b) = (h) ∧ (g) and its characteristic bitstring is σ(h ∧ g) = $\overline{(111000)} \cap \overline{(000001)}$ = (000111) ∩ (111110) = 000110

The set (a) – (h) can be partially represented in the hexagon (DIA), knowing that it constitutes the fragment of a total set of $2^8 = 256$ possible formulas within Σ_2.

(DIA)

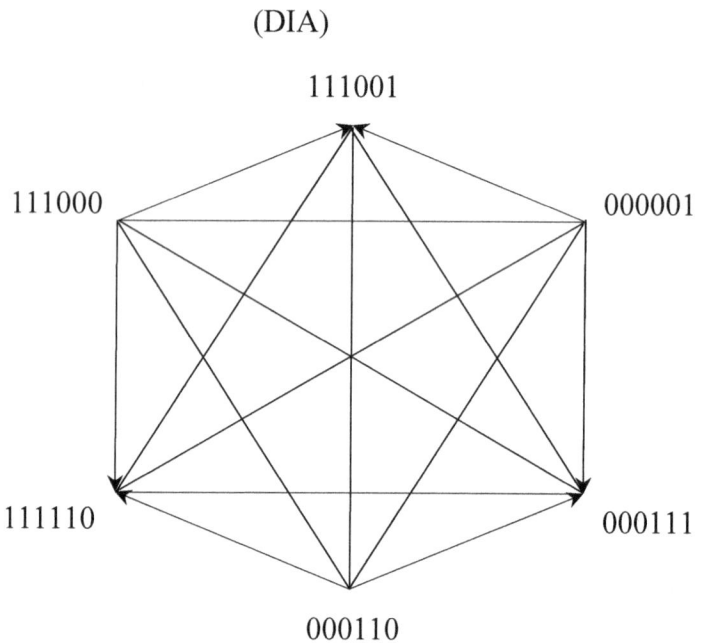

The Boolean calculus also confirms the idea that dogmatism, academicism and scepticism constitute an exhaustive triangle of contraries: *contraries*, because their characteristic bitstrings never overlap with each other in Σ_2 and their intersection is therefore empty: $\alpha \wedge \beta \wedge \varphi = \bot$, *i.e.*

$$\sigma(\alpha) \cap \sigma(\beta) \cap \sigma(\varphi) = 111000 \cap 000001 \cap 000110 = 000000;$$

exhaustive, because the union of the three epistemic attitudes occupies the entire logical space such that $\alpha \vee \beta \vee \varphi = T$, that is to say, σ(α) ∪ σ(β) ∪ σ(φ) = 111000 ∪ 000001 ∪ 000110 = 111111.

This partition semantics can also be applied to von Wright's truth-logics, based on the logical space Σ_3 which characterizes the operators T, F, and T'. Σ_3 turns out to be isomorphic to Σ_1, since Σ_3 also includes three subspaces mentioned in von Wright [29,30]:

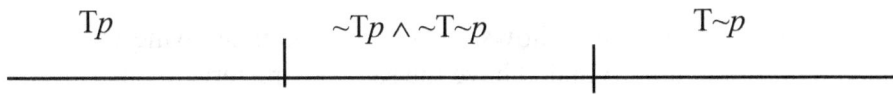

The result is a set of formulas characterized by bitstrings of length $n = 3$, namely:

σ(Tp) = 100; σ(~Tp ∧ ~T~p) = 010; σ(Fp) = σ(T~p) = 001; σ(T'p) = σ(~T~p) = $\overline{σ(T~p)}$ = 110; σ(T'~p) = σ(~Tp) = $\overline{σ(Tp)}$ = 011; σ(Tp ∨ Fp) = 100 ∪ 001 = 101, σ(~Tp ∧ ~Fp) = 011 ∩ 110 = 010.

These valuations match with the logical hexagon (S2) of Woleński [34]:

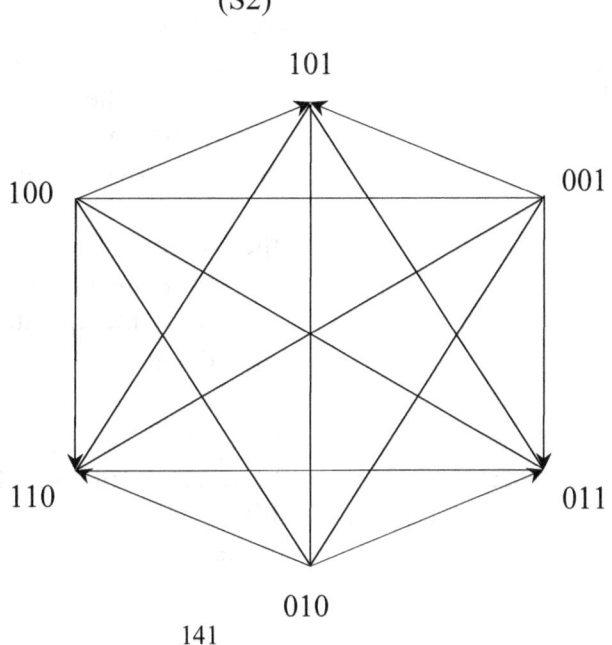

(S2)

We can see here that the two non-unilateral truth values, B and N, confirm what von Wright [29], [30] explained by his metaphor of the transition zone between dry and rainy weather: these are only two ways of expressing the same situation, that is to say, the same logical subspace in Σ_3.

The proof runs as follows, which consists in showing that the two metalogical operators B and N have the same characteristic bitstring:

$$Bp = T'p \ \& \ T'{\sim}p, \text{ so } \sigma(Bp) = 110 \cap 011 = 010$$
$$Np = {\sim}Tp \ \& \ {\sim}T{\sim}p, \text{ so } \sigma(Np) = 011 \cap 110 = 010$$

B and N say 'the same thing', whether in elementary terms of 'strong' or 'liberal' truth. The same conclusion can be reached within $AR_{4[Oi]}$, in order to explain the distinction between the four categories of logical systems according to their interpretation of truth: 'paracomplete' systems give to it a 'strong' *epistemic* meaning, such that 'Tp' means 'p is provable' or 'there is conclusive evidence in favor of p'; 'paraconsistent' systems give a 'liberal' (or 'weak') *epistemic* meaning, such that 'Tp' means 'p is justifiable' or 'there is reason to believe that p (is the case)'; 'normal' (or 'classical') systems give an *ontological* meaning, such that 'Tp' means 'it is a fact (or it is) that p';[18] the paranormal system, finally, combines the three previous interpretations and this explains why no theorem is valid in this system endowed with an absolutely free interpretation.

We come back now to the problem of Naess and Woleński. What should be the correct interpretation of the epistemic attitude of scepticism? This seems to correspond to the operator [A_5] of $AR_{4[Oi]}$, due to its two characteristic partial functions: the first mapping, $T \mapsto \overline{F}$, means that the existence of an argument for p justifies the rejection of its falsehood; the second mapping, $F \mapsto \overline{T}$, means that the existence of an argument against p justifies the rejection of its truth. The values of the domain therefore represent available data, while the values of the counter-domain express the judgment of the speaker determined by such data.

Does the sceptical agent reject the truth of *at least* one proposition, or the truth of any proposition in the sense of a set of sentences K? Woleński explained that only the first interpretation does justice to the contrary attitudes of dogmatism and academism; now we have shown that there are other opposite epistemic attitudes among the expressions (a) – (h), so that nothing seems to prevent the dogmatist from being even more radically opposed to the academician: the first is likely to assert that all truth is knowable, while the second would continue to think that no truth is knowable. The logical question is to ask whether or not dogmatists, academics and sceptics should constitute a set of epistemic attitudes not only exclusive but also, and above all, *exhaustive*. Although Gödel's second theorem of incompleteness gives a strong argument in favor of the interpretation of Naess and Woleński, it is possible to conceive of other more or less radical epistemic attitudes in relation to the three models cited and according to the meaning attributed to the concept of 'truth'.

Examples of seemingly 'irrational' epistemic attitudes come in particular from Indian philosophies, including the Jain theory of *anekantavada* or *saptabhangi* (theory of non-unilateral judgments) and the *catuskoti* (Tetralemma) of the Madhyamika or 'Middle Way' school.[19] These two philosophical stances seem illogical because one seems to accept (the truth of) any sentence while the second would reject them all. In other words, the Jain agent asserts everything and embodies the expression (a), while the Madhyamaka agent asserts nothing. We will limit the examination of this possibility to situations of first-order beliefs, that is to say, to epistemic attitudes bearing on a sentential content and not on themselves (*de se* beliefs). Is it possible to believe the truth of any sentence? This seems to be the case of the sceptic, insofar as he opposes the 'positive' dogmatist who asserts at least one sentence. But since he also opposes the academician by not asserting the falsity of any statement, the sceptic therefore recognizes the truth or falsity of *no* statement. This amounts to a 'non-bivalent' situation in which rejecting the truth of a sentence does not imply asserting its negation, i.e. its falsehood. Although the logic $AR_{[A5]}$ seems to account for this sort of agent, its characteristic matrix does not, however, prevent the assertion of a sentence whenever its assigned truth-value is 'unilateral'.

One can conceive the logic of the sceptic in two ways: either as an attitude of *material* rejection, or as an attitude of *formal* rejection. In the first case, the assertion of a sentence is formally possible but materially impossible, due to the epistemic inability of the sceptical agent to meet the criteria of justification for any sentence. This amounts to making a sort of truncation of the matrices characterizing the sceptic in $AR_{4[A5]}$, such as

p	$[A_5]p$
11	00
■	■
00	00

In the second case, it is formally impossible to assert anything due to the ontological inability of any sentence to meet the criteria of 'strict' truth. This amounts to performing a truncation in the field of truth-values, such that the domain of the sceptic eliminates all assertion and is compelled to interpret any sentence in $V_1 = N = \{00\}$. This situation is mathematically possible, and von Wright mentions it as one of the 16 'truth-logics' resulting from the powerset of the four initial values $B = 11$, $T = 10$, $F = 01$, $N = 00$:

> There are in all 16 different ways in which one can "permit" or "forbid" some or several of the four cases. (We then include the two extreme cases of permitting all four and permitting none of them respectively.)
> These 16 selections answer to 16 different 'truth-logics'. Not all of them seem to be of interest and some of them, moreover, would seem to be identical with one another.
> [30, p. 314.]

The logic of the sceptic would correspond in this perspective to all of the sentences interpreted in the univalent domain $\{N\} = \{00\}$, indicated in red here below:

$$\text{Card}(V_0) = 1 = \{\emptyset\}$$
$$\text{Card}(V_1) = 4 = \{\{11\},\{10\},\{01\},\{00\}\}$$
$$\text{Card}(V_2) = 6 = \{\{11,10\},\{11,01\},\{11,00\},\{10,01\},\{10,00\},\{01,00\}\}$$
$$\text{Card}(V_3) = 4 = \{\{11,10,01\},\{11,10,00\},\{11,01,00\},\{10,01,00\}\}$$
$$\text{Card}(V_4) = 1 = \{11,10,01,00\}$$

As for the epistemic attitude of the Jain, it would correspond to the opposite case (indicated in blue, here above) in which any statement is interpreted within the one-valued domain $\{B\} = \{11\}$. A reason for admitting this formal truncation is given by the *internalist* account of epistemic attitudes in Schang [19], [20], [22]: from his own point of view, the Jain attributes to the concept of truth a 'conventional' meaning (*samvrti-satya*) such that the slightest reason to accept a sentence is sufficient, while the Madhyamakas give to it an 'absolute' meaning (*paramartha-satya*) such that no reason is sufficient to accept any sentence. At the same time, an *externalist* account of epistemic attitudes modifies the domain of valuation of the Jain : his seven conceivable judgments consists in an exhaustive combination of the different kinds of epistemic attitudes which may be either normal and paraconsistent ($v(p) \in \{10,01,00\}$) or normal and paraconsistent ($v(p) \in \{11,10,01\}$).[20]

Admitting such explanations seems essential to prevent the slightest case of assertion. The distinction between assertion and supposition may partially account for these radical epistemic attitudes: the Jain does not assert anything and supposes everything, so that his attitude is more akin to *eclecticism* than optimistic dogmatism; the sceptic asserts nothing and rejects everything, because his criterion of justification is so high that the truth of any sentence must be absolute. The logical effect of these attitudes is such that they cancel out the possibility of a bivalent domain, insofar as any truth-value is *designated* for the Jain and *not designated* for the Madhyamaka. The bi-partition required for the construction of a consequence relation is therefore impossible, and any sentence then turns out logically true or logically false. It is not this path of one-valuedness that Woleński followed to analyze the logic of the skeptic, to whom he attributes a non-assertive and logical behavior at the same time. One way to maintain bivalence consists in replacing the notion of traditional consequence *Cn* by a dual consequence *dCn*, in which consequence does not preserve the truth of sentences but their falsity. Woleński explains the logic of the skeptic in that way, through the attitude of rejection. For all statements A,B:

If *A* is rejected and *B* is a dual consequence of *A*, then *B* is also rejected.[21]

Let us note that, from a sceptical point of view, the concept of dual consequence should be synonymous with preserving *untruth* rather than falsehood (since the falsity of *p* entails the truth of ~*p*). Now the sceptical agent of Naess and Woleński seems to admit classical consequence and still make sense of the attitude of assertion after all. Woleński explains this point as follows:

> Many things concerning rejection can be of course expressed by Cn and negation. For example, the modus tollens leads from asserting $A \Rightarrow B$ and asserting ~*B* (= rejecting *B*) to rejecting *A* (= asserting not-*A*). However, the sceptic does not like the assertion game, even in a mixed form, and certainly he prefers the language that does not commit him to assertion [33, p. 192.]

Does the sceptic take dual consequence to be a mere alternative language game that is equivalent to the assertive language game? It all depends on the interpretation of his attitude towards sentence and the concept of truth. Woleński's version is more 'liberal' than ours, if radical scepticism means that one can attribute the truth to no sentence whatsoever. The dual consequence Woleński deals with is distinct from the traditional relation of consequence, but both are still interchangeable and the discourse of the sceptic does still make sense for a bivalent agent. From our point of view, it is the epistemic attitude $[O_i]$ of the sceptic which is dual with that of the Bivalentists: rejection is untranslatable in terms of assertion, and the discourse of the sceptic is therefore a language game which is meaningless for a bivalent agent. No wonder if it is so difficult to construct a logic characterizing this agent, and the same holds for other agents such as Parmenides, Hegel, or Bradley.

6. Partition Semantics for Non-Suszkian Logics

It is not difficult to construct a logic which does not subscribe to the 'weak' version of PB, when the truth-values are algebraic values reducible to Suszko's logical values. On the other hand, it is much more difficult, if not impossible, to conceive of a logic that does not subscribe to the 'strong' version of (PB'): can we say of a statement that it is designated and not-designated at the same time? If a sentence is true-and-false, it is designated and is not undesignated. If a sentence is neither-true-nor-false, it is undesignated and is not designated. The various responses to Suszko's Thesis, (including [6,12]) did not refute this thesis but advanced alternative kinds of consequence (preserving either falsehood, or untruth). Suszko's Thesis is therefore not 'false' or inconsistent, but it may appear less 'insightful' in the sense that the Tarskian consequence would not be sufficient to understand rationality in a more comprehensive way.

A criterion of insightfulness was proposed by Woleński, in order to show the philosophical irrelevance of the coherence theory of truth: "If a theory is obscure, it should be abandoned; if it does not satisfy its promises, it should also be abandoned, and the same holds for a redundant theory. Since the coherence theory is obscure or it does not satisfy own promises or it is redundant, it should be abandoned" [32, p. 44.] Just as there can be several interpretations of logical principles and epistemic attitudes, Woleński also distinguishes two versions of the coherence theory of truth: a 'mixed' version, which maintains the existence of a true sentence *corresponding to* a fact while defining the truth of the other sentence in terms of coherence (if 'p' is held true in the sense of truth as correspondence, then the disjunction '$p \vee q$' is true because it is coherent with respect to p); a 'pure' or 'Bradleyian' version (with reference to its author, Francis Herbert Bradley), by virtue of which it is a whole system of sentences S which is held true and not the sentences of S. Woleński criticizes this 'pure' definition of truth as coherence because of one main logical defect: the failure of 'down'-compactness, which is the converse of the compactness property and which Woleński defines [32, p. 46] as follows:

> If X is a set of propositions and every finite subset of X is true, then X is also true.[22]

The failure of compactness in the coherence theory of truth is due to the holistic nature of the concept of truth: it is impossible to assign truth to single sentences of S, hence their truth is only 'partial' in the sense that they depend on the truth of all the other sentences of the system. Now this holism is more radical than the holism of the so-called Duhem-Quine thesis, in that it responds to a 'pure' theory of coherence whose meaning is of an ontological order; in contrast, Quine's truth as coherence is a holism of justification, rather than dealing with truth as it stands. Woleński quotes Russell, the main opponent of Bradley's idealism, according to which his doctrine seems obscure because it obeys some 'logic other than ours'.[23]

The alternative is therefore the following, which can be depicted as Woleński's test of insightfulness: either a philosophical theory makes sense, and there is a logic able to explain this theory; or there is no such logic, and the theory does not make sense (it must be rejected, accordingly). It is notably this absence of clearly defined 'logic' which seems to justify the rejection of philosophical theories such as Bradley's therory truth as pure coherence, but also Parmenides' theory of being, Hegel's self-difference (inspired by Heraclitus), or even Heidegger's 'nihilating nothing'. Two questions arise here: Is the 'other logic' Russell was talking about compatible with the standards of modern logic, based on the fundamental relation of consequence? Can a theory be called 'logical' if it does not embed or include any consequence relation? We have seen so far that the plurality of modern logical systems rests on a certain version of logical pluralism, according to which the difference between systems lies in their disagreement about what 'being the case' means.[24]

Now the 'logics' of Bradley, Hegel or Heidegger seem to require more than a pluralism of truth, that is to say, a variety of definitions of the concept of truth within one and the same set-theoretical model (including the 'strict' and 'liberal' truths of von Wright [29], [30]; they seem to require a pluralism of *ontology*, i.e. the construction of models alternative to the mainstream model theory and incompatible with the formal semantics exposed thus far.

Partition Semantics, previously exposed in the analysis of epistemic attitudes, may be able to make sense of some of the "linguistic extravagances"[25] for which it seems impossible to construct one's own logic. Two case studies could appear as cases of 'non-Suszkian logics', i.e. rational systems in which the 'strong' principle of bivalence PB' does not hold: dialectical synthesis, and nothingness.

Several attempts to formalize the Hegelian dialectic have been proposed so far, including da Costa [3] and Rogowski [15].[26] In the former's system $C_0 - C_\omega$, the concept of antinomy is rendered by a 'partial' negation whose applications validate or invalidate PC depending on the structural complexity of the sentences. In the latter's logic of change, a domain of four truth-values is proposed to make sense of the process of 'becoming'. This domain includes 'unilateral' truth-values (the true: 'it is the case only', and the false: 'it is not the case only') and 'non-unilateral' (sub-truth: 'it begins to be the case that', and sub-falsehood: 'it ceases to be the case that'), in order to explain the transition from being to non-being. This logical system partitions the concepts of being and non-being, in the sense that it attempts to explain this continuous transition between these two states in terms of discrete truth-values. This passage takes place through a *cyclical* negation, which turns a 'unilateral' state into a 'non-unilateral' state (and vice versa).[27] However, this system does not seem able to explain the process of dialectical synthesis: it always rests on the concepts of being and non-being, since it explains the concept of becoming as a transition between these two basic states or being and not-being.

A more 'radical' explanation would be to proceed in the reverse sense, without presupposing states and conceiving of being and not being as the results of the process of dialectical synthesis. A model of this kind is proposed in Schang [25], accounting both Bradley's truth as coherence and Hegel's sursumptive' negation.[28] Let x be a kind of initial object, the Absolute, which exhausts the logical space and whose truth-value is the True. The synthesis process is to be interpreted as an object constructor, by successively partitioning this initial exhaustive object into different parts that still 'participate' in it, in the light of the following 'tiered' model:

Level 1

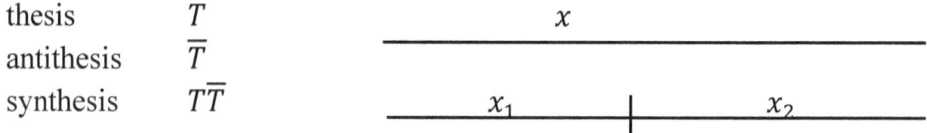

Level 2

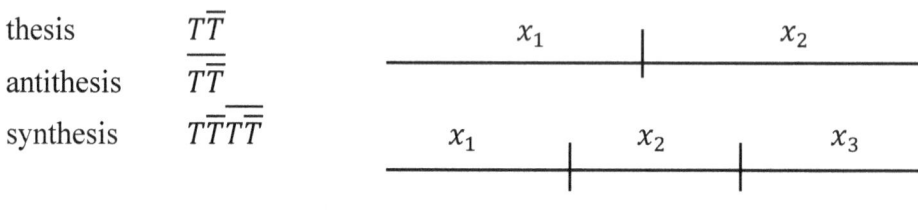

Level 3

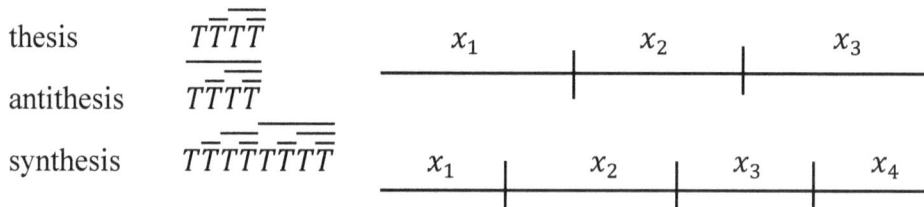

...

The initial being, T, is preserved hereby in each of subsequent states and results from a construction process identical to that of algebraic truth-values: the 'false' $F = \overline{T}$ corresponds to the antithesis of Level 1; the 'true-and-false' corresponds to the synthesis of Level 1, etc. This model is able to explain the meaning of metalogical negation applied to truth-values: it corresponds to the Hegelian negation or *Aufhebung*, which is often translated as a process of 'changing-by-preserving' and which escapes the principle of subsumption with judgments like 'S is P' and 'S is not P'. Hegelian negation thus produces a change by the antithesis, but it guarantees the preservation of the original truth through synthesis. If this model gives meaning to Hegel's dialectic, it shows above all that the Hegelian negation is not an operator applied into a preestablished domain but, rather, a truth-value *constructor*.[29] The same can be said of Heidegger's 'nihilation', which is also not a sentential negation but consists in rejecting a characteristic property of any object. In partition semantics, this means that the 'nihilation' process works like a subtraction operator that decreases the number of bits 1 of the bitstring characterizing an object.[30]

Partition Semantics may also make sense of Bradley's holistic theory of truth as coherence, as well as to the concept of 'nothingness'. If the Hegelian dialectic explains the construction of an *ontology* as an increasing partition of one initial unique model into an increasing number of particular objects, the final set of constructed objects corresponds to Bradley's 'total' or 'absolute' truth T, and each singular object constitutes a 'partial' truth inseparable from T. Conversely, the concept of 'nothingness' designates that which is nothing and cannot be predicated of any object. The length of the bitstring $\sigma(x)$ characterizing any object x makes it possible to distinguish the concepts of relative and absolute nothingness: 'relative nothingness' is an 'object' x such that $\sigma(x) = \bot$ with a number of *finite* bits, while absolute nothingness would be characterized by a logically equivalent bitstring but whose number of bits is *infinite*. This distinction is also found in the constructive process of algebraic truth-values: in a bivalent value domain $2^n = 2^1 = 2$, the 'false' corresponds to the empty set of 'non-true'; in a domain of quadrivalent values $2^n = 2^2 = 4$, it is the 'neither-true-nor-false' which corresponds to the empty set while the 'false' constitutes a proper element of the domain.[31] As for nothingness, it would be this particular 'object' which would remain empty in all the successive domains of truth-values.

Admitting these explanations of truth as coherence, of Hegelian dialectics and of nothingness supposes a certain dose of tolerance with regard to the notion of 'logic': with no relation of consequence, as a process of constructing formal ontologies. In a sense, the relation of consequence rests on a process of discrimination (of the 'true' logic and the 'false' logic) whose philosophical counterpart is the distinction between being and non-being. However, being and non-being presuppose an ontology of stable objects, i.e. substances. A 'Suszkian' logic presupposes in this sense the existence of substances which cannot be reduced to accidental properties, while a 'non-Suszkian' logic does not presuppose any ontology and consists in building models rather than ordering their preexisting components. Bradley's model is not a set of particular objects, but an absolute object that includes everything. Hegel's model includes an absolute object, from which all the particular objects are derived and which participate in it. In contrast to these special models, the logical model is an Aristotelian model: an 'object' is *some*thing, that is to say, a finite set of properties some of which are predicated or not and whose characteristic bitstrings are thus distinguished from any other object in the comprehensive model. We find in this explanation an echo of Aristotle's hylemorphism, according to which every object is a unique combination of form and matter. This 'mixed' ontology contrasts with that of Parmenides and Heraclitus: in the first, everything is a form at rest, so everything 'is' whereas 'becoming' does not make sense; in the second, everything is a matter in movement, so everything 'becomes' whereas 'being' does not make sense.[32] These ontologies therefore involve 'radical' judgments of total acceptance and total rejection, and Partition Semantics is likely to explain what Russell called a logic 'different from ours'. These are non-Suszkian logics, so to speak.

7. Conclusion: What are Truth-Values?

We did not pretend to address here all of Professor Woleński's philosophical and logical writings. However, we hope to have followed the general method of analysis which he has developed so far and which could be depicted as *formal philosophy*: the use of formal tools for the understanding and elucidation of philosophical problems.

The problems discussed here were some logical principles of rationality: PC, PEM, and PB; epistemic attitudes: dogmatism, academism, and scepticism; philosophical theories, such as Bradley's 'pure' theory of truth as coherence and the concept of nothingness. A fundamental tool was used to organize our thoughts on these issues, namely: truth-value, and our main questioning concerned the nature of such an abstract 'object'. Is there a specific answer to this question? Any relativistic response risks reducing logical analysis to an exercise of formal hermeneutics in which the theorist always has a reason to argue and is never at fault. However, this is more or less the answer that we bring to the end of this article, through a certain interpretation of truth-values: these are the *referents* of sentences, in accordance with the first clause FA_1 of FA; but these referents are not reduced to two 'logical' objects which are the true and the false, as opposed to the second clause FA_2. Any response to this subject requires an explanation of the nature of this abstract object.

From the perspective of proof theory, a truth-value means the result of a proof and it does not make sense to assign it to a sentence out of the process of proof. From the perspective of model theory, a truth-value means that a corresponding sentence belongs to a model and it makes no sense to assign a truth-value to it outside any model. The intended referent is therefore either a proof or a membership relation. But it can be even more, if this 'abstract object' of truth-value may receive other formal interpretations.

The partition semantics introduced hereby has attempted to widen the field of interpretations in this way, beyond Suszko's 'logical' values and Łukasiewicz's 'algebraic' values. For Suszko and for Łukasiewicz, truth-values designate classes of sentences that are accepted or rejected and characterize the relation of consequence within a formal logic; in Bradley's theory of truth as coherence or the Hegelian dialectic as we have reconstructed them, truth-values designate classes of objects that differ from the usual sentences of formal logic: it is the totality of sentences, in Bradley's theory; it is no sentence in particular but, rather, an individual object, in Hegel's dialectic. Our conclusion is that the limits of formal logic depend essentially on the meaning attributed to the concept of 'referent'. If truth-values are considered by Frege as proper names, these proper names are very general and can vary in their cardinality: there are only two exclusive according to Frege and Suszko, while there can be more according to Łukasiewicz ; there is an infinite number of inclusive ones, for Hegelian idealists (all included in the 'Great Fact', or the Absolute), while there is none for the Madhyamaka Buddhists.[33] Suszko's situations are also 'truth values' in their own right, once we no longer consider a truth-value as a class intended solely to characterize a relation of consequence. There may be even further interpretations of logical values, such that still relate to the consequence but go beyond the sole area of assertive judgments. An exhaustive treatment of truth-values thus belongs to a broader formal theory of values, but the present paper wanted to stick to the former ones.[34]

The idea of many-valued logics is no 'madness', everything depends on the function assigned to the formal language that makes uses of these. Łukasiewicz's 'madness' may pe pushed even further, as we did hereby. Only Woleński's test of insightfulness can convince us that a theory is not crazy, as long as it is possible to construct an appropriate formal theory of meaning. A formal semantics of partitions purports to fulfill this requirement, just as the semantics of possible worlds did it with respect to the language of modalities.

References

1. Beall, J. C., and G. Restall. *Logical Pluralism*, Oxford: Oxford University Press, 2006.
2. Blanché, R. *Les structures intellectuelles* (*Essai sur l'organisation systématique des concepts*), Paris: Vrin, 1966.
3. da Costa, N. C. On the theory of inconsistent formal systems, *Notre Dame Journal of Formal Logic* 15, 1974, pp. 497-510.
4. Demey, L., and H. Smessaert. Combinatorial Bitstring Semantics for Arbitrary Logical Fragments, *Journal of Philosophical Logic* 47, 2018, pp. 325-363.
5. Englebretsen, G. Knowledge, Negation, and Incompatibility, *Journal of Philosophy* 66, 1969, pp. 580-585.
6. Frankowski, S. Formalization of a plausible inference, *Bulletin of the Section of Logic* 33, 2004, pp. 41-52.
7. Gardies, J.-L. *La logique du temps*, Paris, PUF, 1975.
8. Gauthier, Y. *Hegel: introduction à une lecture critique*, Québec: Presses de l'Université Laval, Coll. 'Logique de la science', 2010.
9. Lemanski, J, and F. Schang. A Bitstring Semantics for Calculus CL, *Logica Universalis*, forthcoming.
10. Łukasiewicz, J. *On the Principle of Contradiction in Aristotle*, Kraków: Polska Akademia Umiejętności, 1910.

11. Łukasiewicz, J. On Three-valued Logic, *Ruch Filozoficzny* 5, 1920, pp. 170-171.
12. Malinowski, G. Q-Consequence Operation, *Reports on Mathematical Logic* 24, pp. 49-59.
13. Piaget, J. *Traité de logique (Essai de logique opératoire)*, Paris: Dunod (1st edition: 1949).
14. Priest, G. The Logic of the Catuskoti, *Comparative Philosophy* 1, 2010, pp. 32-54.
15. Rogowski, S. The Logical Sense of Hegel's Concept of Change and Movement, *Studia Filozoficzne* 6, 1961, pp. 3-39.
16. Schang, F. A Plea for Epistemic Truth: Jaina logic from a Many-Valued Perspective, In A. Schumann (ed.), *Logic in Religious Discourse*, De Gruyter, 2010, pp. 54-83.
17. Schang, F. Negation and Dichotomy, In D. Łukasiewicz and R. Pouivet (eds.), *Scientific Knowledge and Common Knowledge*, Bydgoszcz: Epigram Publishing House/Kazimierz Wielki University Press, 2009, pp. 225-265.
18. Schang, F. Trois paralogismes épistémiques, une logique des énonciations, In P.-E. Bour, M. Rebuschi and L. Rollet (eds.), *Construction. Festschrift for Gerhard Heinzmann*, London: College Publications, 2010, pp. 407-416.
19. Schang, F. Two Indian Dialectical Logics: saptabhaṅgī and catuṣkoṭi, in A. Gupta and J. van Benthem (eds.), *Studies in Logic*: *Logic and Philosophy Today* 29, 2011, pp. 45-74.
20. Schang, F. A One-valued Logic for Non-One-Sidedness, *International Journal of Jaina Studies* (Online) 9 (1), 2013, pp. 1-25.
21. Schang, F. On Negating, In E. Dragalina-Chernaya (ed.), *Ontology of Negativity*, Moscow: Kanon, 2014, pp. 329-376.
22. Schang, F. Eastern Proto-logics, In J.-Y. Béziau, M. Chakraborty and S. Dutta (eds.), *New Directions in Paraconsistent Logics*, New Delhi: Springer Proceedings in Mathematics and Statistics, 2016, pp. 529-552.
23. Schang, F. From Aristotle's Oppositions to Aristotelian Oppositions, In V. Petroff (ed.), *The Legacies of Aristotle as Constitutive Element of European Rationality*, Proceedings of the Moscow International Conference on Aristotle. RAS Institute of Philosophy, October 17-19, 2016, Moscow: Aquilo Press, 2017, pp. 430-445.

24. Schang, F. End of the square?, *South American Journal of Logic* 4 (2), 2018, pp. 1-21.
25. Schang, F. A General Semantics for Logics of Affirmation and Negation, *Journal of Applied Logics* 2020, forthcoming.
26. Schang, F. Question-Answer Semantics, *Revista de Filosofia Moderna e Contemporeâna*, 8, 2020, pp. 73-102.
27. Suszko, R. The Fregean axiom and Polish mathematical logic in the 1920's, *Studia Logica* 36, 1977, pp. 377-380.
28. Turzynski, K. The Temporal Functors in the Directional Logic of Change of Rogowski – Some Results, *Bulletin of the Section of Logic* 19, 1990, pp. 30-32.
29. von Wright, G. H. Truth, Negation and Contradiction, *Synthese* 66, 1986, pp. 3-14.
30. von Wright, G. H. Truth-logics, *Logique et Analyse* 30, 1987, pp. 311-334.
31. Woleński, J. A Note on Scepticism, *Kriterion* 3, 1992, pp. 18-19.
32. Woleński, J. Against Truth as Coherence, *Logic and Logical Philosophy* 4, 1996, pp. 41-51.
33. Woleński, J. Scepticism and Logic, *History and Philosophy of Logic* 1, 1998, pp. 187-194.
34. Woleński, J. The Principle of Bivalence and Suszko's Thesis, *Bulletin of the Section of Logic* 38, 2009, pp. 99-110.
35. Woleński, J. An Abstract Approach to Bivalence, *Logic and Logical Philosophy* 23, 2014, pp. 3-15.
36. Woleński, J. Something, Nothing and Leibniz's Question, Negation in Logic and Metaphysics, *Studies in Logic, Grammar and Rhetoric* 54, 2018, pp. 175-190.
37. Woleński, J. Logical Ideas of Jan Łukasiewicz, *Studia Humana* 8, 2019, pp. 3-7.
38. Zaitsev, D., and Y. Shramko. Bi-facial Truth: a Case for Generalized Truth Values, *Studia Logica* 101, 2013, pp. 1299-1318.

Notes

1. The two versions of bivalence are symbolized PB and PB', in [34]. We will focus here on PB, i.e. the formulation of bivalence in terms of algebraic values.
2. Woleński points that, regarding the psychological interpretation of PC, "Łukasiewicz argues that the last understanding is irrelevant for logic, because it is an empirical fact that people assert contradictory assertions." [37, p. 4]. One might ask two questions about Łukasiewicz's intriguing position with respect to this psychological interpretation of PC. First, why does he believe that the existence of contradictory beliefs does not constitute a sound reason for invalidating PC? Second, are these contradictory beliefs held in the context of transparent or opaque discourse, that is, known or unbeknownst to doxastic agents? Our pragmatic interpretation of PC will take the existence of such contradictory beliefs seriously.
3. 'Sentence' and 'proposition' will be used interchangeably throughout the paper, as they only occur with an indicative use.
4. One could blame this example for confusing what is distinct in the theory of speech acts, namely: assertive acts, and declarative acts. The example of sentence on the metric convention could be considered an example of the latter, and thus show that this sentence is not a proposition. On the other hand, metaphysical propositions are indeed assertive acts and thus confirm Von Wright's view that there are propositions neither true nor false.
5. The author also sees in this liberal interpretation of truth a possible explanation for the process of 'synthesis' in Hegel's dialectic: "I suppose that it is something like that which happened in Dialectical Synthesis". We will return to this process later one, with respect to truth as coherence and nothingness.
6. 'Affirmation' and 'negation' are understood here as illocutionary forces, and not as the locutionary properties of a sentence or propositional content. To avoid confusion between these locutionary and illocutionary aspects, we will only use the phrases 'acceptance' and 'rejection' in the rest of this article.

7. One can also interpret these operators as functions which transform only certain truth-values and leave the others unchanged: [A_1] turns the true into non-false and leaves the false unchanged, for example. They are not 'total functions', in the sense given by Béziau in the Appendix of [35].

8. This means that T constitutes the primary element in the construction of truth value domains: from T comes the false, $F = \overline{T}$, then the other non-bivalent truth-values. We will return to this process of constructing truth values in order to try to shed light on Woleński's reflections on Bradleyian coherence and nothingness.

9. "Having a logic with 2^n logical values, we can always construct its extension with 2^{n+1} logical values" [34, p. 106].

10. The proof of identity of [AN_i] and [NA_i] is provided [25], as well as the redundant operator form of the classical assertion: [AA_i] = [NN_i]. It is also explained that the 'fusion' of operators is distinct from their composition or iteration, of form [A][N]p (acceptance of rejection) and [N][A]p (rejection of acceptance).

11. The details of this general framework will not appear in this paper, due to its irrelevance for the present issue; for a presentation of the syntax and semantics of AR$_{4[Oi]}$, see [25]. The logical constants may be explained as follows in AR$_{4[Oi]}$, for any arbitrary sentences p,q such that their algebraic values are the ordered pairs $v(p) = (X_1,Y_1)$ and $v(q) = (X_2,Y_2)$. Thus: $v(p \wedge q) = (\max(X_1,Y_1), \min(X_2,Y_2))$; $v(p \vee q) = (\min(X_1,Y_1), \max(X_2,Y_2))$; $v(p \to q) = (\max(X_2,Y_2), \min(X_1,Y_2))$.

12. This formation rule means, recalling von Wright's truth-logics, that there are no 'mixed' formulas like [O_i]$p \to p$ in AR$_{4[Oi]}$. Indeed, the expression '[O_i]p' indicates a judgment whereas 'p' indicates a mere sentential content. The formula '[O_i]$p \to p$' is therefore an ill-formed sentence meaning something like 'If the door is closed, then closing the door'). It is because of this syntactic rule that the logical octagon proposed by [34] does not make sense in AR$_{4[Oi]}$, since Woleński admits formulas of types $Tp \to p$ by admitting of sentential variables among its well-formed formulas.

13. Von Wright also states the equivalence of (PEM′) and (2′), due to the Morganian behavior of negation in his paraconsistent truth-logics T′L and T″L: "Is T($p \vee \sim p$) a tautology? The answer is No. It can, in fact, easily

be shown that T($p \vee \sim p$) is logically equivalent in (TL) with T$p \vee$ T$\sim p$, i.e. with the Principle of Bivalence." [29, p. 10.]

14. The T-scheme is of form T$p \leftrightarrow p$, which lies behind the 'deflationary' theory of truth and means that the semantic predicate of truth T adds nothing substantial to the meaning of the sentential content p.

15. Note that the translation of T$p \rightarrow p$ in AR$_{4[Oi]}$ is not [A$_i$]$p \rightarrow p$, which is an ill-formed formula. Rather, it must be rephrased as [A$_i$]([A$_i$]$p \rightarrow p$), 'I accept that everything I accept has an evidence for it'. It turns out that this last formula does not hold with, e.g., [A$_8$]. Indeed, [A$_8$]([A$_8$]11 $\rightarrow$ 11) = [A$_8$](00 $\rightarrow$ 11) = [A$_8$]01 = 01.

16. α = 'I assert that the truth is discoverable'; β = 'I assert that the truth is not discoverable'; γ: 'I do not assert that the truth is not discoverable'; δ = 'I do not assert that the truth is discoverable'; ε = 'I assert that the truth is discoverable or I assert that the truth is not discoverable'; φ = 'I do not assert that the truth is discoverable and I do not assert that the truth is not discoverable'.

17. This abstract object may be a sentence, but also a concept, or even an individual object. See e.g. [9] about the latter case.

18. Any confusion between the 'antirealist' (epistemic) and 'realist' (ontological) interpretations of T risks producing paradoxical consequences if these are admitted within a single, single logical system, which is not the case in AR$_{4[Oi]}$. This seems to be the case with the 'Fitch Paradox', whose conclusion is that a proposition is true if and only if it is known: $p \rightarrow$ Kp. The 'paradoxical' consequence of this antirealistic definition of truth is indeed based on a 'mixed' formal language in which p and Kp belong to the same object language. A syntactic criticism of this paradox is formulated in [18], which consists in refusing any mixed formula as an ill-formed formula (thus blocking the initial premise of the paradox). Another anti-paradox strategy appears in the 'bi-facial' system [38], which consists in distinguishing two kinds of truth-values: ontological (T and F), and epistemic (1 and 0).

19. See in particular [16], for a many-valued analysis of *saptabhangi* in either 7- or 15-valued domains. See also [19,20] for a 1-valued (therefore non-Suszkian) analysis of *saptabhangi* and *catuskoti*.

20. The cardinal of the Jain *seven* judgments follows from combinations of different epistemic attitudes, which yields this general model of

particular models or valuations: $\{\{10\},\{01\},\{11\},\{10,01\},\{10,11\},\{01,11\},\{10,01,11\}\}$, in the normal and paraconsistent system J_{7G}; $\{\{10\},\{01\},\{00\},\{10,01\},\{10,00\},\{01,00\},\{10,01,00\}\}$, in the normal and paracomplete system J_{7M}. Thus, there are $2^3 = 8 - 1$ possible ways of judging any sentence from a set of 3 single epistemic attitudes, the 8th forbidden case being the one in which sentences are neither accepted nor rejected. We take this last situation to match with the Madhyamaka stance of 'silence' or peace of mind, such that the sentence is entertained without being judged at all. See [14] about this interpretation which seems to corresponds to the above special case $\{\varnothing\}$ of von Wright's 16 truth-logics.

21. For example, let $A = p$ and $B = p \wedge q$.

22. Woleński specifies that the principle of compactness trivially holds in the Bradley system, since this principle is expressed in the form of a conditional whose antecedent is false. It is only the converse of this principle that is awkward.

23. "The coherence-theory is generally advocated [...] in the connection with logic entirely different from ours." [32, p. 45.]

24. The *pluralism* of the criteria for assigning truth is defended in particular in [1]; it is opposed to Carnap's logical *relativism*, where the disagreement does not come from the meaning of truth but from the meaning of logical constants (regardless of their truth conditions).

25. "Perhaps Heidegger's and Sartre's linguistic extravancies, like 'nihilation' or 'neantization' well illustrate various troubles with the (absolute) Nothingness." [36, p. 187.]

26. For a discussion of Rogowski's logic of change, see especially [28] and also [7], [17]. This logic modifies the previous explanation given by von Wright [29], [30] about the drizzle, which he presented as a case of rain and no rain and which becomes hereby a case of 'sub-falsehood'.

27. This cyclical negation cannot be translated in $AR_{4[Oi]}$, because it establishes between truth-values an ordering relation which does not correspond to any of the rejection operators $[N_i]$.

28. The concept of 'sursumption' was created by Gauthier [8] to point out the idea that Hegel's being overhangs (and *includes*) contradictory

qualities, as opposed to the principle of subsumption that rules contradictory (and *exclusive*) judgments of form 'S is P' and 'S is not P'.

29. This operator is compared to the succession operator S of Peano's arithmetic such as $S(n) = n + 1$.

30. This operator may be viewed as a precedence operator P dual to S, such as $S(n) = n - 1$.

31. The relativity of nothingness is evoked by Woleński with the example of the silent composition of John Cage, *4'33*. Woleński poses the following question: "Let us assume that every year Cage would have written a piece of finite length, but always a minute longer than the present one. Would then the structure of, say, *6'33* be the same as that of *4'33*?" [36, p. 187]. Our answer is No: the two compositions would have been different, due to the difference in length n in their characteristic bitstrings.

32. The distinction between 'nothing', 'something' and 'everything' is explained in [23] as a difference between their respective bitstrings: 'something is some thing' and 'no thing is nothing' hold, whereas 'every thing is everything' does not.

33. The constructive process of truth-values that leads to various domains of valuation shows this increasing process of relative bitstrings, where every finite bitstring relates to a special kind of proper name: a *Kripkean* proper name, which behaves as a uniquely identifying expression.

34. 'Good' and 'wrong' may also occur as the referents of *moral propositions*, i.e. expressive speech-acts by means of which Leo Strauss' *Reductio at Hitlerum* is rendered as a moral version of Modus Tollens. See Schang, F., "Moral Inferences" (draft) and "Political Oppositions" (talk to be delivered at the next 7th World Congress on the Square of Opposition, Leuven, September 7-11, 2021).

Reism, Concretism and Schopenhauer Diagrams

Jens Lemanski

University of Hagen
Universitätsstr. 33
58084 Hagen, Germany
e-mail: jens.lemanski@fernuni-hagen.de

Michał Dobrzański

University of Warsaw
Krakowskie Przedmieście 3 Street
00-927 Warsaw, Poland
e-mail: michaldobrzanski@uw.edu.pl

1. Introduction

In his article published in *The Stanford Encyclopedia of Philosophy* on the doctrine of reism, Jan Woleński remarks that it has been anticipated by a number of philosophers from antiquity to modernity. The list includes names such as Thomas Hobbes, Gottfried Wilhelm Leibniz, and Franz Brentano, and eventually points at the Polish philosopher Tadeusz Kotarbiński as the one who has presented the "most developed version" of the doctrine [31]. Kazimierz Ajdukiewicz and Woleński concretize the abstract concept of reism by dividing it into an ontological (only things are real) and a semantic dimension (concepts must be reduced to things) [1], [31]. In this paper, we argue (1) that the above-given list should be enhanced by the name of the philosopher Arthur Schopenhauer, who was born in Danzig in 1788 and died in Frankfurt in 1860, and who is for example known for having influenced Wittgenstein [18], [7]. Moreover, we argue not only for reism in Schopenhauer's work but also for the fact (2) that in his *Berlin Lectures* of the 1820s Schopenhauer has developed a diagrammatic method of concretization.

Argument (1) may seem quite unexpected, given the fact that Schopenhauer is known as a thinker who holds that the whole world is a manifestation of a metaphysical and irrational will [30, p. 34] – a stance that seems to be nowhere less than at complete odds with e.g. Kotabiński's reist program. To prove this not fully adequate, we will focus in Section 2 on Schopenhauer's methodology and offer a reading of it which gives strong foundations for viewing him as a reist. In this section, we will also reconstruct the most important elements of his philosophy of language of his *Berlin Lectures* as, until recently, they have not drawn much attention among scholars.

Argument (2) is addressed in Section 3. Here, we will develop a diagrammatic method that Schopenhauer used in his *Berlin Lectures* to illustrate his reistic doctrines. For Schopenhauer, logic diagrams are the best way to concretise what can normally only be expressed in abstract terms. Therefore, we argue that they can show another, namely diagrammatic dimension to understand the position of reism or concretism. These diagrams have already been introduced in [8] as a general tool for philosophy of language. Although the diagrammatic method has certain similarities to the diagram systems of e.g. Leonhard Euler, Immanuel Kant, and even John Venn, we use the term "Schopenhauer diagrams" to avoid further clarifying the relationship to already known logic diagrams.

2. Schopenhauer's Reist Philosophy of Language

In this section, we will first give an introduction to Schopenhauer's philosophy of language (2.1), then present his theory of concepts (2.2), and finally argue that Schopenhauer's theory can be called reistic (2.3). In this presentation (2.1 – 2.2) and argumentation (2.3), we refer mainly to the writings from Schopenhauer's Berlin period (1818 – 1830) and especially to his *Berlin Lectures*.

2.1. Introduction to Schopenhauer's Philosophy of Language

§1 State of Research

Despite the claim of Jan Garewicz, the Polish translator of, among others, *The World as Will and Representation* (*WWR*) that Arthur Schopenhauer's philosophy "has found a strong resonance in the period of scientism and positivism" [10, p. 32], the German philosopher's work on philosophy of language and logic seems to remain almost unknown to the researchers currently concerned with these topics. This might be somehow connected with the fact that it is in the manuscripts for his *Berlin Lectures* [23], [24], written in the 1820s, that he dedicates his attention to these issues in the most systematic and profound way. The lectures were until recently[1] only available in an edition published over 100 years ago, during the ending of a period which might be considered the peak of interest for his philosophy[2] [3, p. 13 f.]. However, it is not that Schopenhauer does not work on these topics in his other works. In fact, the topics of language and concepts appear in his writings throughout his career, starting from his dissertation (1813) until his final work *Parerga and Paralipomena* (1851), and seem to constitute an object of his reoccurring philosophical interest which plays an important role for his philosophical system [6, pp. 11-12].

§2 Hierarchies of Language

In a recent paper, Matthias Koßler argued that Schopenhauer's theory of language cannot be simply reduced to a nominalist, instrumental theory, in which language is treated as a tool for describing empirical objects. However, Koßler admits at the end of his paper that "[n]evertheless Schopenhauer talks about language as a tool […]" and adds: "He [sc. Schopenhauer] does not reject these aspects of language but places them into a hierarchic order of different uses of language" [15, p. 23]. Without further discussion on whether the instrumental theory of language is the core or just one of several uses of language distinguished by Schopenhauer, it certainly is present in his analysis of language and, significantly for our purpose, it provides a framework which seems to concur with reism[3].

§3 Language within Schopenhauer's System

As the titles of his main work (*The World as Will and Representation* = *WWR*) and the more detailed *Berlin Lectures* (*The Doctrine of the Essence of the World and the Human Spirit*) suggest, Schopenhauer assumes that there are only two ways of knowing the world that can be attributed to humans – as representation and as will [25, p. 129], [24, p. 41]. Whereas the parts of his writings in which he discusses the world as will can be, broadly speaking, interpreted as the presentation of his metaphysics, the examination of the world as representation contains elements of his epistemology and methodology. Not surprisingly, Schopenhauer in both works quite early in the presentation of his system already discusses the problem of language and specifically the possibilities of application of concepts for the description of intuitive and mental facts. This discussion can be found in the rather short paragraph 9 of the first volume of *WWR* (about 10 pages long) and is then significantly enhanced in the notes for Schopenhauer's *Berlin Lectures*, which encompass more than 100 pages on language and logic.

§4 Idealism and Empiricism

The starting point for the construction of Schopenhauer's system seems quite paradoxical. On the one hand, he assumes the Kantian, idealistic view that the "being of things is identical with their cognition" [Das Seyn der Dinge ist identisch mit ihrem Erkanntwerden] [23, p. 113], which he expresses in his claim that all the world is our *representation* (i.e. the world that we perceive is not the thing-in-itself). On the other hand, Schopenhauer sees the framework of the phenomenal world with its *a priori* forms of cognition as somehow the natural way of knowing the world[4] and the only possible foundation for any further philosophical and metaphysical investigations. He opposes any possibility of deducing the truth about the world from reason alone and instead makes the claim that any metaphysics should be founded upon the immanent experience of the subject or even „empirical sources of knowledge" [23, p. 152], cf. also [14, p. 363]. Thus, Schopenhauer simultaneously assumes (1) the idealist stance that empirical reality is a creation of the subject's cognition and (2) the empiricist distinction of empirical sources and the subject's knowledge. This is possible because he treats the empiricist dualism as the starting point for the construction of a philosophical system, which eventually is monist.

§5 Ontological and Epistemological Interpretation

Consequently, the distinction of empirical sources and the subject's knowledge should not be interpreted ontologically, but epistemologically. Schopenhauer does not claim that what is empirical is ultimately real. He only claims that we experience the subject-dependent phenomenal world as having two dimensions, namely intuitive objects and abstract thoughts, and this is the outlook we need to assume as the starting point for philosophical reflection, as from it we get out data for the investigation of the world. We need to do so, even if we are philosophically aware of the idealistic character of human cognition.

2.2. Schopenhauer's Theory of Concepts

§6 Two Classes of Phenomena

According to §5, philosophical reflection sets off with considering the world as representation or a collection of representations (phenomena). These phenomena can be grouped into two classes: (1) intuitive and (2) abstract phenomena. The character of this classification is epistemological, as the reason for it is provided by the different modes of cognition of both classes of representation: (1) intuitive representations are recognized by understanding [Verstand] [23, p. 207], (2) the abstract ones by reason [Vernunft] alone [23, p. 242]. "All our representations", Schopenhauer says, "can generally be divided into visual [anschauliche] and merely thought-like [gedachte], intuitive [intuitiv] and abstract, into images and concepts" [23, p. 118]. As can be seen, this distinction is also equated with the differentiation of phenomena into "images" (which can be "seen") and – significantly! – "concepts" (which can be "thought of"). Obviously, this must lead Schopenhauer to provide a solution to such questions as the characteristics of these two classes and their mutual relation.

§7 Intuitions

From a systematic point of view, intuitive phenomena are contrary to abstract phenomena. That is, if something is an intuitive phenomenon, it cannot also be an abstract phenomenon and *vice versa*. From a historical point of view, Schopenhauer dissociates from the theories of mere sensory data of ancient and modern rationalists and empiricists and adopts a reduced Kantian theory of intuition: the intuitive phenomena provide the *material* data which we can then express in terms of concepts. However, the reception of this data is conditioned by the *form* of space, time, and causality [23, p. 57, cf. also pp. 146, 172], which allows us to experience, i.e. to absorb sensory data. Therefore, space, time, and causality are *a priori* valid and they generate the *hic et nunc* of intuitive representation. In the end, it seems plausible to assume that Schopenhauer understands intuitive representations as reality [Wirklichkeit] [23, p. 207] which is empirical and gives immediate, direct knowledge. However, we need not forget that this dualism between intuitive and abstract phenomena is only epistemological, but not ontological.

§8 Concepts

Concepts, the second class of phenomena, are characterized as "a very peculiar class of representations that exist alone in the human mind" and which are "*toto genere* different" from intuitive representations. This difference is expressed above all in the fact that concepts can only be thought of abstractly, but not observed in intuitive representation [23, p. 242]. In other words: concepts are not empirical, intuitive objects, but they are experienced by the subject as something like – using modern terminology – mental states. Furthermore, Schopenhauer holds that "every concept as a general, not a specific, representation has what is called a sphere, a circumference", which refers to a set of objects (both other concepts as well as real objects, see below) that can be conceived by it [23, p. 257].

§9 Abstraction and Concept

How, then, are concepts made? Reason produces concepts by abstracting from the many properties of objects that are given in intuitive representation: The concept therefore contains less than the [intuitive, JL&MD] representation itself"; it is created by "seeing away from what is unique in the individual [Wegsehn vom Besondern der Individuen]" [23, pp. 249, 252]. Thus, a concept "does not contain everything" that is given or contained in its intuitive basis. Because of this "innumerable intuitive objects" can be thought of with the help of a concept [23, p. 249]. On the basis of an intuitive representation an abstract, mental reconstruction of it can be formed, which is generally applicable to many other objects in intuitive representation. This generalization, which consists of the liberation from the *hic et nunc* of intuitive representation (§7), thus enables the mental grasp of abstract, past and future facts, and these in turn can become human motives for action.

§10 Generality of Concepts

Schopenhauer points out that the general applicability of concepts for intuitive representation is not the result of the process of development of concepts – i.e. abstraction from one or many intuitive objects or concepts (§9) – but it is a result of their substantial nature, i.e. their being merely mental, which is characterized by the absence of temporal-spatial determinations. It is, therefore, possible and even necessary that a concept that has arisen by abstracting from properties of one single intuitively given object can potentially be applied to several objects [23, p. 256]. Schopenhauer says: "a concept is always general, even if there is only one thing that is thought by it; and only a singular intuition that gives it content, is a proof of it" [23, p. 276 f.].

§11 Classes of Concepts

Concepts, as abstract representations or thoughts, are also divided by Schopenhauer into two general classes, *concreta* and *abstracta*. *Concreta* are abstracted directly from intuitive representations, and *abstracta* are formed by abstracting from some characteristics of "concepts or genera [Gattungen]". According to his examples, *concreta* are for instance red, dog, house, and *abstracta*: color, relation, friendship. He strongly reiterates that this classification is, strictly speaking, inauthentic or wrong, because all concepts are in fact abstract and only "what is intuitive is actually concrete" [23, p. 252]. By using the (inauthentic) terms *concretum* and *abstractum* he seems to refer to the original Latin meaning, where *abstrahere* stands for "taking away" (cf. Schopenhauer's claim that all concepts are an effect of a "seeing away" above) and *concrescere* for "growing together". The classes are only helpful in understanding the relation of concepts to the empirical world. Schopenhauer uses an allegory: if we think all concepts that we have as a building, then the ground on which it stands will be intuitive representations, the ground floor will be *concreta* and the higher floors will be *abstracta* [23, p. 252]. The more general a concept is, the further away it is from empirical reality.

§12 Intuition-Concept-Hierarchy

By reference to the classes of concepts (§11), Schopenhauer claims an epistemological hierarchy, in which intuitive objects (§7) *precede* concepts as a source of knowledge. He also denies any kind of innatism, i.e. the presence of *a priori* concepts in the human mind: "the whole abstract faculty of reason [sc. the conceptual] is a secondary one, which presupposes intuition" [23, p. 235]. The dependence of concepts on intuitive representations is a consequence of how he understands the process of the development of concepts, namely as "reproduction, repetition, of the archetypal intuitively given world" [23, p. 251]. Consequently, concepts become dependent on intuitive reality as the source of information or data that they contain (§9). This finds its expression for instance in the following quotation: "the whole world of reflection [...] rests on the intuitive one as its basis of cognition" [23, p. 252]. This is the reason, why Schopenhauer repeatedly refers to concepts as "representations of representations" [23, p. 249]. Concepts have meaning only in relation to empirical reality[5] and the more abstract a concept is, the less meaning it has.[6]

2.3. Schopenhauer's Reism

§13 Reism

This leads to the core claims of what could be called Schopenhauer's reism. As has been shown, within his basic idealistic outlook (§4) he develops a theory of two types of cognition, intuitive and conceptual (§§6-8), and puts them into an epistemological hierarchy (§12), as he holds that concepts have meaning only in reference to empirical, intuitive objects, without which they would be nothing. But even more crucially, he also holds that concepts can be understood if and only if they can be referred back to intuitions. For a concept to be distinct and meaningful [deutlich], it must be possible to fill it with empirical content. The "common explanation that the concept is distinct if it can be broken down into its characteristics is not enough" as long as these characteristics cannot be traced back to intuitive representation, i.e. to clear perceptions [23, p. 254 f.]. Schopenhauer concludes: "From our entire inquiry it has become evident to everyone that the origin of all knowledge and the foundation of all science lies in direct knowledge, that is, in intuition. Intuition is the last source of all truth: all abstractions, all concepts, are only substitutes and only for their other use, are they the substance of our knowledge; their truth is always an indirect one: the source of all evidence is intuition. All knowledge, all thinking, which does not eventually lead to some kind of intuition, is empty" [23, p. 539].

§14 Reist Language Criticism

For Schopenhauer, we only have meaningful [deutliche] concepts if we are able to replace our abstract concepts with references to intuitive reality. It follows that we should be able to break *abstracta* down to *concreta*, so that *concreta* refer [hindeuten] to empirical reality [cf. 23, p. 254 f.]. This idea is also one of the foundations, if not the most important one, of his repeated criticism of Scholastics and German idealists, whose proponents are criticized for their abundant use of very abstract concepts [32]:

> "Especially in philosophy, the danger is great that one rises so high from abstraction to abstraction that the way back to intuitive phenomena [Rückweg zum Anschaulichen] is no longer to be found: then the whole knowledge is empty: one operates with mere concepts that are no longer based on intuition: such knowledge is like paper-money that cannot be cashed anywhere"[7] [23, p. 539].

Obviously, Schopenhauer is criticizing here the improper use of language, and the problem, which he refers to, is that these philosophers' terminology does not allow a clear reference to reality[8] [4]. Putting it into reist terms: such abstract terminology cannot be reistically translated.

§15 Kotarbinski's Reism

The stance that abstract concepts need to be broken down into *concreta*, which again can be referred back to intuition is strongly reminiscent of what Kotarbiński says about how a reist should proceed: "for every declarative sentence (statement) that includes abstract terms he tries to find an equisignificant statement including no such terms". Also, definitions of abstract and concrete terms are provided: "By abstract terms I mean here all those which are not concrete, and by concrete I mean all, and only those, terms which are names of things" [16, p. 441]. This formulation of the reist program is almost identical to Schopenhauer's language criticism and even uses similar terminology. However, one important difference should be pointed out. In Kotarbiński's reism *concreta* are "names of things". This seems to at least presume the ontological statement that the world, which we conceptualize in language, consists of things. Indeed, soon after its presentation, Kotarbiński's reism was subject to a debate regarding its interpretation as either (1) the ontological claim that "every object is a thing" or (2) a semantic program which states that "every 'name' which is not a name of a thing" should be held "for an apparent name" [1, p. 610 f.]. It was also pointed out that ontological reism uses abstract terms for expressing its main theses and in consequence should be disqualified according to its own rules [31]. After this criticism, Kotarbiński himself reformulated his reism into a semantic, normative program to free language from abstract names for clarity [31].

§16 Between Ontological and Semantic Reism

If we try to consider Schopenhauer's language criticism according to this classification, it seems obvious that his postulate that we should be able to concretize *abstracta* and eventually refer concepts to intuitions can be interpreted as a semantic program (2) which formulates criteria how language should be used. It can indeed be understood as something quite similar to Kotarbiński's semantic reism, as a "program with the aim of thorough 'de-hypostatization' of humanities" [or better: philosophy], with the goal of "turning it into a discipline which uses clear, simpler and more comprehensible language, even if less 'sublime' or 'deep'" [33, p. 564 f.]. The question of whether Schopenhauer could be interpreted also as an ontological reist is more complex, given his steady claim about the idealistic character of representation (§4), which from a transcendental point of view denies the existence of things.

§17 Epistemological Concretism

For this reason, the distinction into ontological and semantic reism seems not appropriate for analyzing Schopenhauer's reism. In fact, the crucial problem is that whereas in Kotarbiński's reism *concreta* are "names of things", Schopenhauer understands them as direct abstractions from intuitions (§13). It has to be underlined at this point that the original term for intuition which he uses is *Anschauungen*, which in German strongly connotes visuality, as can be seen in §6, where abstract concepts are confronted with "images". This reiterates the fact that his understanding of the distinction into intuitions and concepts is epistemological and not ontological (§5). Thus it seems plausible to leave out the ontological question and interpret his semantic reism from §16 as an epistemological claim, which could be reformulated as follows: "in order to be meaningful, abstract concepts have to be replaceable with concepts which can be intuited [or better in this context: *visualized*, 'veranschaulichen']". Or more simply: *in order to understand concepts we need to visualize them*. This is strongly founded upon Schopenhauer's axiomatic claim that all new knowledge lies in intuition [Anschauung] (§12) and that only intuition is truly concrete (§11). For this reason, the term "reism" seems inadequate and it is more suitable to refer to Schopenhauer's doctrine as "concretism" – a term, which Kotarbiński used synonymously with "reism". However, it should be specified that this is an epistemological, not an ontological concretism, cf. [31].

§18 Visualization

To sum up, with recourse to [23, pp. 251-256], one could define the following claims of Schopenhauer's epistemological concretism: (1) only the objects in intuitive representation are concrete; i.e. language is always abstract and only those terms are called (inauthentic) *concreta* that directly correspond to concrete intuition (§11). (2) If concepts are to be meaningful [deutlich], it must be possible to break them down into increasingly concrete concepts (§13), so that one can finally use these concrete concepts to indicate or to point to intuitive phenomena [hindeuten]. It follows that in order to make concepts comprehensible, we need a theory of visualization. And indeed, Schopenhauer makes several attempts to provide such theories for different fields. He does so e.g. for mathematics (cf. his visualization of the Pythagorean theorem, which he holds to be self-explanatory [23, p. 425]) or for poetry [24, p. 317], but it is for the visualization of concepts and language that he formulates the most developed theory in the *Lectures*. This theory is based on diagrams, which we discuss in Section 3.

3. Schopenhauer Diagrams and Epistemological Concretism

In this section, we will first give a short introduction to Schopenhauer diagrams (3.1), then develop a so-called level theory for concretism (3.2), with the help of which we can finally provide a tool for Schopenhauer's epistemological concretism, a semantic program in many ways similar to reism, in form of intuitive diagrams (3.3).

3.1. An Introduction to Schopenhauer Diagrams

§19 Schopenhauer's Diagrams

In his *Berlin Lectures*, Schopenhauer develops a diagrammatic logic that can be used to illustrate semantic positions, topics, and problems. The diagrams that Schopenhauer uses in his treatises on language, logic, and eristic are for him the most important method of concretizing abstract topics since diagrams intuitively illustrate what can only be formulated by using abstract concepts or signs [19]. For Schopenhauer, even abstract algebraic or conceptual theories of mathematics and logic must always be based on an intuitive representation that has an isomorphism to certain diagrams. Although Schopenhauer explains the function of logic diagrams in more detail [20], [5], he does not give precise rules for their application in philosophy of language. In what follows, we will sketch a theory of Schopenhauer diagrams based on four main principles (CI, PI, CE, PE) with which two diagrams given in Schopenhauer's philosophy of language (Fig. 1 and 2) can be analyzed and further developed.

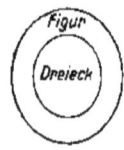 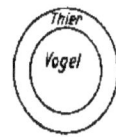 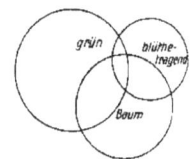

Fig. 1 (PL I, 258): Figur = figure; Dreieck = triangle; Thier = animal; Vogel = bird

Fig. 2 (PL I, 257): grün = green; blüthetragend = flower-bearing; Baum = tree

§20 Complete Sphere Inclusion (CI)

Let us assume that in Fig. 1 we see a diagram that shows at least four terms in the form of four spheres. Two concepts are assigned to a CI, which is shown as a subset ($\subseteq$)[9] in the diagram: (**CI-1**) The sphere that denotes the concept *triangle* is completely contained within the sphere of the concept *figure*, i.e. *triangle* $\subseteq$ *figure*. (**CI-2**) The circle denoting the concept *bird* is completely contained within the sphere of the concept *animal*, i.e. *bird* $\subseteq$ *animal*. In Fig. 2 we find no representation of CI.

§21 Partial Sphere Inclusion (PI)

PIs exist when two spheres have an intersection ($\cap$) in the diagram. In Fig. 1 we find two PIs, since the two larger spheres are partially contained in the smaller spheres: (**PI-1**) The concept *figure* is partly contained in the sphere of *triangle*, i.e. *figure* $\cap$ *triangle*. (**PI-2**) The concept *animal* is partly contained in the sphere of *bird*, i.e. *animal* $\cap$ *bird*. In Fig. 2 we find even more PIs: (**PI-3**) The sphere that denotes the concept *tree* partially intersects the sphere of the concept *green*, i.e. *tree* $\cap$ *green*. (**PI-4**) Also *green* and *flower-bearing* intersect, i.e. *green* $\cap$ *flower-bearing*, and (**PI-5**) *flower-bearing* and *tree*, i.e. *flower-bearing* $\cap$ *tree*. Furthermore, we see in Fig. 2 that PI can also occur with more than two terms, since (**PI-6**) the sphere of the concepts *green*, *tree* and *flower-bearing* intersect in such a way that there is a common intersection in the middle of the diagram, i.e. *green ∩ tree ∩ flower-bearing*.

§22 Complete Sphere Exclusion (CE)

However, Fig. 1 also shows that two of the four spheres with the other two remaining spheres show neither CIs nor PIs (∆): (**CE-1**) The sphere of the concept *figure* has neither CIs nor PIs with *animal*, i.e. *figure ∆ animal*; (**CE-2**) Due to (CE-1), (CI-1) and (CI-2) must also apply that *triangle* and *bird* possess neither CIs nor PIs, i.e. *triangle ∆ bird*. From (CE-1) and (CE-2) it is now also evident that one of the larger spheres with one of the smaller spheres has neither CIs nor PIs, i.e. (**CE-3**) *figure ∆ bird* and (**CE-4**) *animal ∆ triangle*.

§23 Partial Sphere Exclusion (PE)

PEs are present when CIs or PIs exist between two conceptual spheres, but a relative complement (\) remains that is not described by CIs or PIs between these two concepts. In Fig. 1 we find two PEs, namely where the inside of the larger sphere is not covered by the smaller one, i.e. (**PE-1**) *figure \ triangle* and (**PE-2**) *animal \ bird*. Since PIs were found in Fig. 2, we see here three PEs with two concepts: (**PE-3**) The sphere denoting the concept *tree* does partially not intersect the sphere of *green*, i.e. *tree \ green*. (**PE-4**) Also *green* and *flower-bearing*, i.e. *green \ flower-bearing*, and (**PE-5**) *flower-bearing* and *tree*, i.e. *flower-bearing \ tree*. If one thinks about the union (∪) of all three spheres and subtracts (**PI-6**) from it, the result is one of several possible PE ratios including three concepts, i.e. (**PE-6**) (*green ∪ tree ∪ flower-bearing*) \ (*green ∩ tree ∩ flower-bearing*).

§24 Relations

Based on §§2-5 we can already establish some relations for the individual principles: For CI it is *transitive*, so that for all spheres x, y, z applies: If $CIxy$ and $CIyz$, then $CIxz$. For PI it holds that it is *symmetrical* so that for all spheres x, y holds: $PIxy$ implies $PIyx$. Also, CE is *symmetric*, so for all spheres x, y: $CExy$ implies $CEyx$. For PE it is *not symmetric*, because for some spheres x, y is valid (e.g. PE-1, PE-2): If $PExy$, then not-$PEyx$.

§25 Regions and Frames

Concept development normally starts with only one sphere of a *concretum* (e.g. *bird*, *animal*), but in relation to other spheres they form new ones (e.g. PE-1: *animal* \ *bird*). This is done by the four principles that form different regions (R) inside and outside a given conceptual sphere. In order to understand this concept formation more precisely, however, it is first necessary to examine the syntax of the respective diagrams with regard to the specific regions. These regions are marked in the diagrams D1 and D2, which structurally correspond to Figs. 1 and 2. To make it clear exactly what belongs to a diagram and what does not, we place a square frame (F) around the diagram.

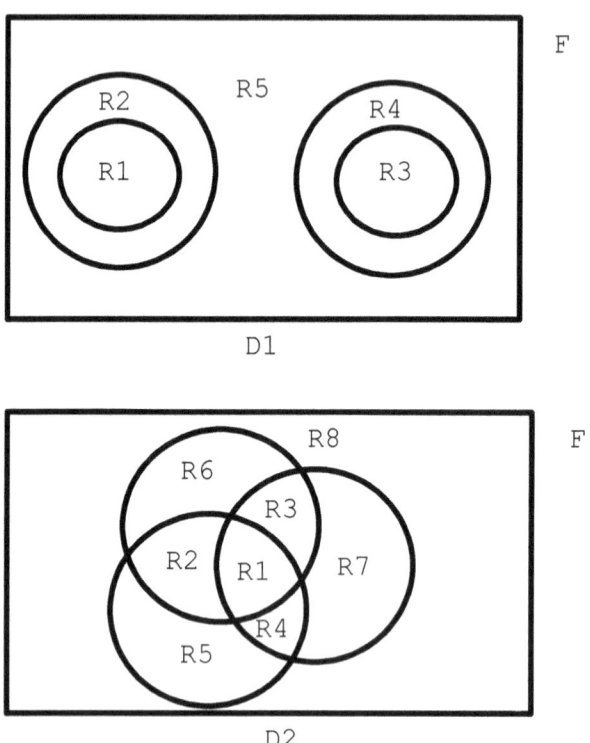

§26 Semantics of Regions

If we transfer the semantic meanings that we have gained in Fig. 1 and 2 with the help of the four principles to the syntactic designations of the regions in D1 and D2, we can make the following assignments.

For Fig. 1 and D1: (CI-1) = {R1}; (CI-2) = {R3}; (PI-1) = {R2}; (PI-2) = {R4}; (PE-1) = {R2}; (PE-2) = {R4}. For Fig. 2 and D2 it applies: (PI-3) = {R1, R2}; (PI-4) = {R1, R3}; (PI-5) = {R1, R4}; (PI-6) = {R1}; (PE-3) = {R4, R5}; (PE-4) = {R6, R2}; (PE-5) = {R3, R6}; {R5} in D1 must also be present, otherwise (CE-1) and (CE-2) could not be displayed. But if we assume {R5} in D1, we must also consider {R8} in D2 to be useful, since both are constructed according to the PE principle: (*figure* ∪ *animal*) Δ F ={R5} in D1; (*tree* ∪ *green* ∪ *flower-bearing*) Δ F = {R8} in D2.

3.2. A Level Theory for Concretism

§27 Abstracta and Concreta

For Schopenhauer, concepts are not uniform; rather, he distinguishes concepts into different levels, which are classified according to the degree of abstraction or concretion. As described in §11, the reference to various levels is justified by the allegory of the building: Terms with different degrees of abstraction are assigned to different levels of the building. Although all terms are abstract, they can be divided (inauthentically) into *abstracta* and *concreta*. Since we will see below that the division into *abstracta* and *concreta* is too imprecise, we add a level degree for concepts Cs, in short: C-level, which is determined by the number of abstraction steps: *1st level C, 2nd level C, n level C.*

§28 Law of Reciprocity

Each concept has a certain circumference and content [23, p. 258]. From a modern point of view, one can call the circumference the extension and the content the intension. Extension and intension of a concept (C_{Ext}, C_{Int}) stand thereby in an inverse relationship: The larger the extension of a concept, the smaller the intension and *vice versa*. If, for example, C_{Ext} can be described by a natural number x of a sequence from 0 to n ($[0,n] := \{x \in \mathbb{N}_0 | 0 \leq x \leq n\}$), then $f(x) = n - x$ applies to C_{Int}. This relationship can be called the Law of Reciprocity, which became prominent through Kantian logic [11], [21]. If the number of C-level is known, then a suitable quantity can be given for n with the following formula:

n = number of C-levels − 1. Let us take the following example: If we set the number of C-levels = 6, then $n = 5$. Furthermore, $C_{Ext} = 5$, if $C_{Int} = 0$. If $C_{Ext} = 4$, then $C_{Int} = 1$, etc.

§29 Building Scheme

With the building allegory given in §§11, 27 we can now set up a scheme (Fig. 3) that illustrates the example of a Law of Reciprocity with $n = 5$ given in §28. Due to the lack of space, the scheme is abbreviated between *2nd level C* and *6th level C*, as indicated by the dotted arrows. Here *3rd level C* ($C_{Ext} = 2$ and $C_{Int} = 3$), *4th level C* ($C_{Ext} = 3$ and $C_{Int} = 2$), *5th level C* ($C_{Ext} = 4$ and $C_{Int} = 1$) are missing. At the very bottom is the object that is given in intuitive representation. All C levels are abstractions from intuitive representation. Therefore concepts are also called abstract representations or representations of representations (§§11, 12).

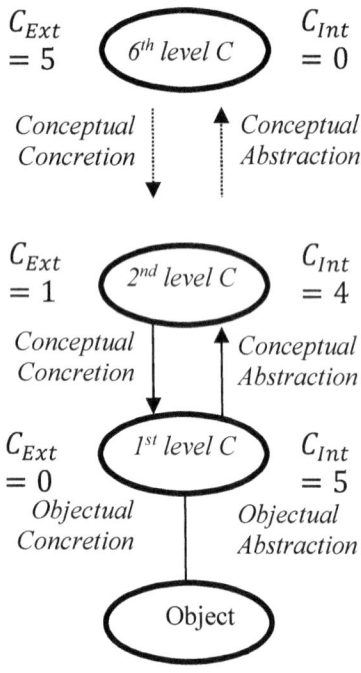

Fig. 3 Building Scheme

§30 Abstraction and Concretion

We see in the building scheme (Fig. 3) that between each level processes of abstraction and concretization take place. If one takes up the modern distinction [9], [29] between *objectual* and *conceptual abstraction* (or concretion), one can also make a corresponding classification of processes, as can be seen in Fig. 3. Only for conceptual abstraction and concretion applies the Law of Reciprocity (§28): If conceptual concretion takes place, C loses a degree of extension but gains a degree of intension. In the case of conceptual abstraction, C gains a degree of extension but loses a degree of intension. Note that the sequence from 0 to n (§28) is a degree and does not indicate the actual number of given objects. Since concepts are always general (§10), we can only indicate the degree of the relation between C_{Ext} and C_{Int}, but never the exact number of possible objects designated by C.

§31 Designations of C-levels

By the building scheme (Fig. 3) it is well recognized that *being-abstract* and *being-concrete* are in most cases relative designations: A term has a relative abstraction and concretion if it has a C-level above and a term below it. For example, a *2nd level C* is *more abstract* compared to *1st level C*, but *more concrete* compared to the *3rd level C*. In such cases we speak of *Abstract-Concrete Concepts* or *ACC* for short. In our example (§§28 et seq.) *1st* and *5th level C* are no ACCs, because they have no C-level either below or above. Thus, we can call a *1st level C* as a *Bottom-Level Concretum* or *BLC* and *5th level C* a *Top-Level Abstractum* or *TLA*. These designations cannot only be justified diagrammatically but also by using the degrees of C_{Ext} and C_{Int}. For TLA, $C_{Ext} = n$ and $C_{Int} = 0$; for BLC, $C_{Ext} = 0$ and $C_{Int} = n$; and for all ACC, C_{Ext} and C_{Int} must be > 0 and $< n$.

§32 Concept and Object

According to §§9, 30, a concept is an objectual abstraction of certain objects given in intuitive representation. According to the building scheme (§26), this definition applies directly to a BLC or *1st level C*, while all other concepts on a higher C-level are abstractions from the lower C-levels, i.e. conceptual abstractions (§30). A concretion of a concept at a higher C-level (ACC and TLA) can therefore only be achieved by its reduction to a BLC or *1st level C*. What this concretion of BLCs might look like, however, is only indicated in Schopenhauer's work: One can say that that objectual concretion is made for instance through deictic references ("Hindeuten", §14) accompanying speech acts.[10] For example, pointing to a certain object when using BLCs such as *red*, *dog*, or *house* (§11) may be an act of concretion. Anyway, we have represented objectual concretion and abstraction by a simple line in Fig. 3 to illustrate the difference to conceptual concretion or abstraction illustrated by arrows.

3.3. Concretion of Concretism with Schopenhauer Diagrams

§33 A Level Theory for Schopenhauer Diagrams

But how can the level theory established in Section 3.2 be applied to the Schopenhauer diagrams outlined in Section 3.1? A key to this attempt of making concretism more concrete with the help of a diagrammatic dimension is to focus on the etymological meaning of abstraction and concretion (§11) and its isomorphism with the four principles of Schopenhauer diagrams, i.e. PI, CI, PE, CE (§§ 20-23). In the following, we assume that the I-principles PI and CI correspond to concretion, but the E-principles PE and CE to abstraction. This can be seen in the design of Schopenhauer diagrams since in the case of I-principles spheres grow together (*concrescere*), whereas in the case of E-principles they are subtracted from each other (*abstrahere*).

§34 Definitions

We now use the Law of Reciprocity (§§12, 28) and say: The more a region (§§25-26) is restricted by I-principles (∩, ⊆), the higher is the degree of intension (C_{Int}) and the more concrete is the concept. But the more a region is defined by E-principles (∆, \), the higher the degree of extension (C_{Ext}) and the more abstract the concept. We further define that the C-principles have a higher concretion (CI) or abstraction (CE) than the P-principles (PI, PC), if the concepts determined by them are related in one diagram.

§35 First Example: D1

In D1, according to §26, we find five regions that can be described by all four principles. By referring to §22, we see that the regions {R1}, {R2}, {R3}, and {R4} are in a balanced CE ratio: Each of these four regions is completely excluded from two others. Thus, for {R1}, {R2}, {R3} and {R4}, the level degree cannot be determined by CE. According to §26, this does not apply to {R5}: Since {R5} = (*figure* ∪ *animal*) ∆ F and since *triangle* ⊆ *figure* (CI-1) and *bird* ⊆ *animal* (CI-2), according to the transitivity-relation of CI (§24) it applies that {R5} = (*triangle* ∪ *bird*) ∆ F. Thus {R5} is completely excluded from all other conceptual spheres. {R1} and {R3} must be considered as *1st level C* or BLC according to the definitions given in

§34 since they are the only conceptual spheres to which CI principles can be applied (see CI-1 and CI-2 above). For {R2} and {R4}, they partly exclude and partly include terms, i.e. (PI-1) = {R2}; (PI-2) = {R4}; (PE-1) = {R2}; (PE-2) = {R4} (§23).

§36 Evaluation of D1

Let us summarize the results of §35. For {R5} is completely excluded from all other conceptual spheres, {R2} and {R4} are partially included, partially excluded, but {R1} and {R3} are completely included, then applies: {R5} = TLA (3rd *level C*), {R2} and {R4} = ACCs (2nd *level C*), {R1} and {R3} = BLCs (1st *level C*). So since D1 denotes 3 *C*-levels, it makes sense to set $n = 2$ (§25) and determine that for {R5} $C_{Ext} = 2$ and $C_{Int} = 0$, {R2} as well as {R4} $C_{Ext} = 1$ and $C_{Int} = 1$, and {R1} as well as {R3} $C_{Ext} = 0$ and $C_{Int} = 2$ applies.

§37 Second Example: D2

According to §26, we find eight regions in D2 that can be described by three principles, i.e. PI, PE, and CE. Furthermore, §26 says that the only CE region is {R8}, which is excluded from all other regions, i.e. (*tree* ∪ *green* ∪ *flower-bearing*) ∆ *F*. The regions {R5}, {R6}, {R7} are each formed by two PE and one PI, e.g. {R6} = (*green* \ *tree*) ∩ (*green* \ *flower-bearing*). The regions {R2}, {R3} and {R4} are each formed by one PI and one PE, e.g. {R2} = (*green* ∩ *tree*) \ *flower-bearing*. {R1}, however, is constructed without E-principles, only by PI, e.g. *green* ∩ *tree* ∩ *flower-bearing*.

§38 Evaluation of D2

Let us summarize the results of §37. For {R8} is completely excluded from all other spheres of concepts, {R5}, {R6} and {R7} are partly included, partly excluded, but {R1} is partly included by all spheres, then applies: {R8} = TLA (4th *level C*), {R5}, {R6} and {R7} = ACC (3rd *level C*), {R2}, {R3} and {R4} = ACC (2nd *level C*) and {R1} = BLC (1st *level C*). So since D2 denotes 4 *C*-levels, it makes sense to set $n = 3$ (§28) and determine that for {R8} $C_{Ext} = 3$ and $C_{Int} = 0$, for {R5}, {R6} and {R7} $C_{Ext} = 2$ and $C_{Int} = 1$, for {R2}, {R3} and {R4} $C_{Ext} = 1$ and $C_{Int} = 2$ and for {R1} $C_{Ext} = 0$ and $C_{Int} = 3$ applies.

§39 Concretization

According to §13, there must be a way back to intuitive phenomena in D1 and D2 if concepts are meaningful [deutlich]. In D1 this means a way back to the two BLCs, either {R1} or {R3}. In D2 a reduction to {R1} is required. For {R1} in D1, for example, we can say that it is a BLC to which not only the concept *figure* but also *triangle* applies. In {R1} in D2 we can say that the BLC designates an object that can be described with the expressions *green*, *tree*, and *flower-bearing*. All terms or regions in D1 and D2 which are connected with at least one BLC by an I-principle can be traced back.

§40 Top-Level Abstracta

However, TLAs cannot be traced back to BLCs as they are associated with all other terms by the CE-principle. TLAs are therefore characterized by the fact that they are negations of all other terms that are marked in a diagram. From {R8} in D2, for example, we know that it denotes all objects that are not green, not a tree, and not flower-bearing. The amount of objects that it denotes is immeasurable, especially when compared to the objects that are trees, or that are trees and bear flowers, etc. But other than *non-tree*, *non-green* and *non-flower-bearing*, we know nothing of {R8} in D2. For Schopenhauer, these TLA are not meaningful [deutlich], since there are no positive characteristics. Its extension is very high, but its intension is completely low. Because of the only negative relation to all other concepts in the diagram, a TLA can therefore not be traced back to a *concretum*, BLC or intuitive representation. According to the reistic criterion (§14) TLAs are therefore only confused [verworren], [23, p. 255] or meaningless words.

4. Summary and Outlook

In Section 2, we have presented Schopenhauer's philosophy of language and in particular his theory of concepts as given in the *Berlin Lectures*. It has been shown that Schopenhauer's theory of concepts can be described as reistic in the widest sense: Without intuitive representations, there would be no abstract representations, so all meaningful *abstracta* must be reduced to *concreta*, which indicate to intuitive representations. Reism itself, however, is a concept that remains abstract if it is not concretized, as e.g. Jan Woleński does, by pointing out an ontological and a semantic dimension. For Schopenhauer's theory, however, the distinction into ontological and semantic reism seems not appropriate. Rather, it seems to make sense to call his approach epistemic concretism due to the role of *concreta* and their relationship to concrete representation. 'Concretism', however, is a term, which Kotarbiński used synonymously with 'reism' therefore the choice of words to describe Schopenhauer's theory plays only a minor role. Much more important is that Schopenhauer introduces a further dimension that helps to understand his reistic or concretistic philosophy of language: Schopenhauer uses diagrams to concretize the degrees of abstraction and concretion of concepts and their relationship to the intuitive representation. We have introduced and discussed this diagrammatic dimension of his philosophy of language in Section 3.

However, research on Schopenhauer's philosophy of language and Schopenhauer diagrams is still in its infancy: As already indicated in §§15 et seq., for example, we have not yet been able to elaborate on all dimensions involved in Schopenhauer's reism. A more precise attempt at clarification, which we cannot undertake in this paper, would have to discuss, for example, the role of phantasm as a possible reference point of *concreta* (§32), but also take into account Schopenhauer's idealistic-transcendental philosophical position with regard to intuitive phenomena. Furthermore, we have reduced Schopenhauer's philosophy of language here to an instrumental theory (§2). We have also ignored certain contextualist approaches in Schopenhauer's *Berlin Lectures*.

However, in connection with Schopenhauer's concretism, there are many more historical and systematic questions for future research: Largely unexplained is Schopenhauer's influence on the philosophers and logicians of the early 20th century mentioned in §1. Furthermore, it can be assumed that Schopenhauer's philosophy of language could be made clearer in a critical comparison with other prominent reists such as Brentano or Kotarbiński. Furthermore, the question remains open whether Schopenhauer's criterion of reist language philosophy also does justice to the controversial concepts of his own theory, e.g. the will, Platonic idea, etc.

Finally, research on Schopenhauer's logic diagrams is also in its infancy: Since Schopenhauer formulated principles of diagram use mainly for the theory of judgement, but not for the philosophy of language, other further interpretations, developments, and applications of his diagrams are conceivable. Of course, the results presented here should also be applied to more complex diagrams that have more than four spheres and where all principles are involved. Furthermore, the question arises as to the relationship of Schopenhauer diagrams to historical ones, e.g. Euler, Kant, Krause, Venn, Peirce diagrams, or to modern systems of diagrams in semantics or logic. This raises the question of which 'observable advantages' Schopenhauer diagrams have and which principles and notations are best suited to describe them [28], [2]. In this paper, however, it was our sole aim to show Schopenhauer's reistic position in his *Berlin Lectures* and its concretization through Schopenhauer diagrams.

References

1. Ajdukiewicz, K. ,Elementy teorii poznania' Tadeusza Kotarbińskiego, In T. Kotarbiński, *Elementy teori poznania, logiki formalnej i metodologii nauk,* Wrocław, Warszawa, Kraków: Zakład Narodowy imienia Ossolińskich, 1961, pp. 607-631.
2. Bellucci, F. Observational Advantages: A Philosophical Discussion, In P. Chapman, G. Stapleton, A. Moktefi, S. Perez-Kriz and F. Bellucci (eds.), *Diagrammatic Representation and Inference. Diagrams 2018.* Lecture Notes in Computer Science (vol. 10871), Cham: Springer, 2018, pp. 330-335.
3. Beiser, F. C. *Weltschmerz. Pessimism in German Philosophy, 1860-1900*, Oxford: University Press, 2016.
4. Birnbacher, D. Schopenhauer und die Tradition der Sprachkritik, *Schopenhauer-Jahrbuch* (99), 2018, 37-56.
5. Demey, L. From Euler Diagrams in Schopenhauer to Aristotelian Diagrams in Logical Geometry, In J. Lemanski (ed.), *Language, Logic, and Mathematics in Schopenhauer*, Basel: Birkhäuser (Springer), 2020, pp. 181–206.
6. Dobrzański, M. *Begriff und Methode bei Arthur Schopenhauer*, Würzburg: Königshausen & Neumann, 2017.
7. Dobrzański, M. Problems in Reconstructing Schopenhauer's Theory of Meaning: With Reference to His Influence on Wittgenstein,

In J. Lemanski. (ed.), *Language, Logic, and Mathematics in Schopenhauer*, Basel: Birkhäuser (Springer), 2020, pp. 25-45.

8. Dobrzański, M., and J. Lemanski. Schopenhauer Diagrams for Conceptual Analysis, In A.-V. Pietarinen, *Diagrammatic Representation and Inference. 11th International Conference, Diagrams 2020 Tallinn, Estonia, August 24–28, 2020,* Lecture Notes in Artificial Intelligence (vol. 12169), Cham: Springer, 2020, pp. 281-288.

9. Fine, K. *The Limits of Abstraction*, Oxford: Clarendon Press, 2002.

10. Garewicz, J. *Schopenhauer*, Warszawa: Wiedza Powszechna, 1988.

11. Hauswald, R. Umfangslogik und analytisches Urteil bei Kant, *Kant-Studien* (101), 2010, pp. 283-308.

12. Juhos, B. *Inwieweit ist Schopenhauer der Kantischen Ethik gerecht geworden?* Wien: Phil. Diss., 1926.

13. Kleszcz, R. Criticism and Rationality in the Lvov-Warsaw School, In D. Kubok (ed.), *Thinking Critically: What Does It Mean? The Tradition of Philosophical Criticism and Its Forms in the European History of Ideas*, Berlin, Boston: de Gruyter, 2018, pp. 161-172.

14. Koßler, M. Die eine Anschauung – der eine Gedanke. Zur Systemfrage bei Fichte und Schopenhauer, In L. Hühn (ed.), *Die Ethik Arthur Schopenhauers im Ausgang vom Deutschen Idealismus (Fichte/Schelling)*, Würzburg: Ergon, 2006, pp. 349-364.

15. Koßler, M. Language as an 'Indispensable Tool and Organ' of Reason: Intuition, Concept and Word in Schopenhauer, In J. Lemanski (ed.), *Language, Logic. and Mathematics in Schopenhauer*, Basel: Birkhäuser (Springer), 2020, pp. 15-24.

16. Kotarbiński, T. Reism: Issues and Prospects, *Logique et Analyse* 11 (44), 1968, pp. 441-458.

17. Kotarbiński, T. Przedmowa, In A. Schopenhauer, *Erystyka czyli sztuka prowadzenia sporów*, Kraków: Wydawnictwo Literackie, 1973.

18. Lemanski, J. Schopenhauers Gebrauchstheorie der Bedeutung und das Kontextprinzip: Eine Parallele zu Wittgensteins ›Philosophischen Untersuchungen‹, *Schopenhauer Jahrbuch* (97), 2016, pp. 171-195.

19. Lemanski, J. Means or End? On the Valuation of Logic Diagrams, *Logiko-filosofskie studii* (14), 2016, pp. 98–122.

20. Moktefi, A. Schopenhauer's Eulerian diagrams, In J. Lemanski (ed.), *Language, Logic, and Mathematics in Schopenhauer,* Basel: Birkhäuser (Springer), 2020, pp. 111-128.

21. McLaughlin, P., and O. Schlaudt. Kant's Antinomies of Pure Reason and the 'Hexagon of Predicate Negation', *Logica Universalis* (14), 2020, pp. 51-67.
22. Schlick, M. Nietzsche und Schopenhauer, In M. Schlick (ed. Iven, M.), *Gesamtausgabe: Abteilung II: Nachgelassene Schriften*, vol. 5.1, Wien, New York: Springer, 2013.
23. Schopenhauer, A. *Philosophische Vorlesungen. Vol. I*, In A. Schopenhauer (ed. P. Deussen and F. Mockrauer), *Schopenhauers sämtliche Werke. Vol. IX.*, München: Piper, 1913.
24. Schopenhauer, A. *Philosophische Vorlesungen. Vol. II*, In A. Schopenhauer (ed. P. Deussen and F. Mockrauer), *Schopenhauers sämtliche Werke. Vol. X.*, München: Piper, 1913.
25. Schopenhauer, A. *The World as Will and Representation. Vol. 1.*, translated and edited by Judith Norman, Alistair Welchman and Christopher Janaway, Cambridge: Cambridge University Press, 2010.
26. Schopenhauer, A. *Der handschriftliche Nachlass: Bd. III. Berliner Manuskripte (1818-1830)*, ed. by A. Hübscher, Frankfurt a.M.: W. Kramer, 1970.
27. Schopenhauer, A. *Parerga and Paralipomena. Vol. 2.*, transl. by A. Del Caro, C. Janaway, Cambridge: Cambridge University Press, 2015.
28. Stapleton, G., Jamnik, M., and A. Shimojima. What Makes an Effective Representation of Information: A Formal Account of Observational Advantages, *Journal of Logic, Language and Information* 26 (5), 2017, pp. 143-177.
29. Tennant, N. A General Theory of Abstraction Operators, *The Philosophical Quarterly* 54 (214), 2004, pp. 105-133.
30. Woleński, J. The History of Epistemology, In I. Niiniluoto, M. Sintonen and J. Wolenski (eds.) *Handbook of Epistemology*, Dordrecht: Kluwer Academic, 2004, pp. 3-54.
31. Woleński, J. Reism, In E. N. Zalta (ed.), *The Stanford Encyclopedia of Philosophy* (Summer 2012 Edition), URL: http://plato.stanford.edu/archives/sum2012/entries/reism/ (11.05.2020).
32. Xhignesse, M.-A. Schopenhauer's Perceptive Invective, In J. Lemanski (ed.), *Language, Logic, and Mathematics in Schopenhauer*, Basel: Birkhäuser (Springer), 2020, pp. 95-109.
33. Zaręba, M. Reizm Tadeusza Kotarbińskiego a prakseologiczna koncepcja sprawstwa, *Przegląd Filozoficzny – Nowa Seria* 3 (83), 2012, pp. 559-575.

Notes

---

1. A slightly modified re-print was published by Volker Spierling in 1984ff. A new edition of the lectures by Daniel Schubbe is currently being published at Felix Meiner Verlag. The publication of the part containing Schopenhauer's considerations on language and logic is currently scheduled for December 2020. An English translation does not yet exist.
2. Beiser points out that the interest in Schopenhauer peaked between the years 1860 and 1914. Significantly, this is also a period in which the founding texts of modern philosophy of language appear. Whether there is any relation between these two facts, however, needs further examination, even if it has already been pointed out that Schopenhauer's philosophy had an impact on Wittgenstein [18], and there is an obvious reception of Schopenhauer in Logical Positivism (e.g. Béla Juhos wrote his PhD-thesis on Schopenhauer [12] and Moritz Schlick lectures on Schopenhauer [22]) and in the Lvov-Warsaw School (e.g. Schopenhauer was quoted at various texts of Kazimierz Twardowski and Kotarbiński wrote the introduction to the Polish translation of Schopenhauer's *Eristic Dialectic* [17]).
3. In this respect, Kotarbiński is very precise: "Thus it is obvious that reism, or concretism, is a variation of nominalism" [16, p. 442].
4. In his *Lectures* he states for example that this is the way of knowing the world by the "philosophically crude" people, who have not yet philosophically reflected upon the world [23, p. 463].
5. In his Berlin period, Schopenhauer found the term "natural education" for this, by which he postulated that empirical experience precede abstract knowledge [26, p. 260; 27, pp. 562-563].
6. For a more detailed explanation of this, encompassing some terminological problems of Schopenhauer's theory, see [7, p. 33 ff.]
7. It has to be pointed out here that Kotarbiński uses a very similar allegory of paper-money in reference to abstract concepts and their role in the reist outlook: "Every banknote, cheque, and promissory note must be exchangeable into gold on demand, which does not mean that all payments are made in gold" [16, p. 444].
8. Interestingly enough, the founder of the Lvov-Warsaw School, Kazimierz Twardowski, also formulated such criticism of German Idealism [13, p. 162].

9. Schopenhauer diagrams are not diagrams of set theory, but nevertheless the notation of set theory is suitable for describing Schopenhauer diagrams. In contrast to naïve set theory, however, we normally assign only one principle, and thus one set-theoretical sign, to each relation of two diagrammatic elements. A detailed study of the notation of Schopenhauer diagrams is planned for the future.

10. Schopenhauer assumes that there are also other possibilities, e.g. though phantasms [6, p. 43 ff.].

Deontic Relationship in the Context of Jan Woleński's Metaethical Naturalism

Tomasz Jarmużek

Nicolaus Copernicus University in Toruń
Stanisława Moniuszki 16/20 Street
87-100 Toruń, Poland
e-mail: jarmuzek@umk.pl

Mateusz Klonowski

Nicolaus Copernicus University in Toruń
Stanisława Moniuszki 16/20 Street
87-100 Toruń, Poland
e-mail: mateusz.klonowski@umk.pl

Rafał Palczewski

Nicolaus Copernicus University in Toruń
Stanisława Moniuszki 16/20 Street
87-100 Toruń, Poland
e-mail: rafal.palczewski@umk.pl

1. Introduction

Relating deontic logic is a deontic logic that introduces an additional condition about relating the formulas with the normative system into semantics. Such logic allows for an extensive range of philosophical considerations, as it does not clearly define what a normative system is, and how to informally understand the so-named evaluation of connection. In this work we will show that this gap can be filled by referring to the metaethics of Jan Woleński. We will learn that both the relating deontic logic – through a certain response to the so-called Jörgensen's Dilemma – as well as Jan Woleński's metaethics, which, where it draws on the Standard Deontic Logic (SDL), is affected by its problems; benefit from the above.[1]

We will begin with a brief presentation of SDL and its fundamental problems associated with individual theses or rules. Subsequently, we will show how relating deontic logic allows us to avoid these problems. Then, we will outline Woleński's metaethical stance, in order to combine it with informal aspects of relating deontic logic in the last part of the paper.

The primary objective of the paper is to indicate the effectiveness of combining two independent stances: logical and metaethical. In the paper, we limit ourselves merely to deontic logic, to the normative concepts analysed herein, while omitting what is also the subject of Jan Woleński's analyses and also find formal representations (often very close to deontic ones), that is, imperative and bonitive sentences, or more broadly: axiological ones. At the same time, we omit many formal details related to the relating deontic logic, or more broadly to the relating logic as such, see [9], [10], [12].

2. The Standard Deontic Logic and its Problems[2]

In SDL, the modal concepts of obligation and permission correspond to the alethic concepts of necessity and possibility, respectively. The element that distinguishes SDL within the family of all modal logics is the validity of the axiom (D). The standard model of the semantics of possible worlds for deontic logic takes the following form:

$$<W, Q, v>$$

where W is a non-empty set of possible worlds, Q is a serial relation of accessibility between the worlds, and v is a classical valuation of propositional variables in the possible worlds. Hilpinen [5, p. 163] describes the possibility of deontic interpretation of such a model in the following way:

> [...] the "standard semantics" [i.e. possible worlds semantics] of deontic logic [...] gives an intuitively plausible account of the meanings of simple deontic sentences when the deontic alternatives to a given world u are taken to be worlds (or situations) in which everything that is obligatory at u is the case; they are worlds in which all obligations are fulfilled. Hence, the worlds related to a given world u by R [accessibility relation, authors] may be termed deontically perfect or ideal worlds (relative to u).

According to Hintikka [7, p. 189], deontic alternatives are different possible variations of the initial world, where the deontic values, required from the perspective of some normative system, occur simultaneously. "These deontic alternatives are also "deontically perfect worlds" of sorts: all obligations, both these that obtain in the actual world and those that would obtain in such an alternative possible world, are assumed to be fulfilled in each of them."

Consequently, what is obligatory must occur in all such worlds; whereas, what is permitted must occur in at least one.

However, let us point out that in deontic alternatives, the sentences that are not obligations in the given normative system, may be true. Thus, there are sentences that do not express obligations, but carry some deontically neutral content. So, how to distinguish those sentences that are true and express obligations from the ones that are true but carry deontically neutral content? Moreover – as we well know – the standard approach leads to various paradoxes, such as the Ross paradox, the good Samaritan paradox, or the paradoxes of derived obligation, extensively described in the literature on the subject, see [1, pp. 268-270], [5, pp. 163-167], [6, pp. 58-64]. Some of them, as described by Carmo and Jones [1, p. 268], result from the closure of the obligation operator O under the logical consequence relation. "The first group of paradoxes has its origin in the closure of the O-operator under logical consequence (that is, in the fact that SDL, like any normal modal logic, contains the (RM)–rule: if ⊢ $A \to B$, then ⊢ $OA \to OB$."

Another problem is closure under the Necessitation Rule, that results in any logical truth expressing obligation in each deontic situation. Following Carmo and Jones [1, p. 270], it can be stated that: "A second problem of SDL has do the with the O-necessitation rule itself, according to which any tautology (more generally, any theorem) is obligatory, which is incompatible with the idea that obligations should be possible to fulfil and possible to violate."

By all means, the closure under the Necessitation Rule in combination with axiom (K) classical logic and the Detachment Rule, allows for deriving the (RM)–rule. Thus, it allows us to obtain the same paradoxes as due to the (RM)–rule. The possibility to create an obligation from each logical truth is also strange because the laws of logic may not remain related to the given normative system whose perspective we are aiming to consider. Logical truth need not be obligatory, nor logical false prohibited, since, from the perspective of the given normative system, they can be completely non-relevant. That is to say, SDL allows for too wide an approach to obligation, prohibition and permission.

3. Relating Deontic Logic

Relating deontic logic is based on the empirical observation that any sentence that is obligatory, prohibited or permitted, is such from the perspective of some value system, or, to put it simply, a normative system. Thus, from the empirical point of view, there are no absolute obligations, nor absolute permissions. Thus, when referring to an obligation, prohibition or permission, we always do so with regard to some value system which orders, permits or prohibits.

The above observation leads to the conclusion that the sentences that do not remain related to the considered normative system, express neither obligation nor prohibition – their content is simply neutral. On the other hand, the sentences that are neutral in relation to the given normative system state what is undoubtedly allowed by the given system, for they cannot express prohibition. Similarly, no sentence that is obligatory from the perspective of a given normative system can carry neutral content since a normative system does not prohibit anything it is not related to. In order to take into account on the formal ground the above-given observations, we complement the conventionally defined semantics of deontic logic with a new element, that is, a family of subsets of a set of formulas:

$$\{R_w\}_{w \in W},$$

thus obtaining the following ordered quadruple

$$<W, Q, v, \{R_w\}_{w \in W}>.$$

Consequently, for every possible world, we determine a subset of formulas, thus representing the fact that, in the given world, the given sentences are related to the given normative system. This relation can be generally termed as the deontic relationship.

In the universe of varying possible worlds, some sentences may become deontically related, while others may cease to be such. The quality of being deontically related can be understood as a deontic relevance, which is an opposite to being neutral with respect to the given normative system. Hence, as we can see, in our semantics there is no direct representation of a normative system, instead we take into account its perspective by differentiating two sentences expressing what is, and, respectively, what is not related to the system.

The introduction of the deontic relationship's representation into the model results in a substantial change in the truth-conditions for deontic sentences. In the proposed approach, what a given sentence states is obligatory, provided that in all deontic alternatives, the sentence is true and remains related to the given normative system, which is as follows:

$$w \vDash OA \text{ iff for all } u \in W, \text{ if } Q(w,u), \text{ then } u \vDash A \text{ and } A \in R_u.$$

Whereas, what a given sentence states is permitted if it is true in some deontic alternative, or is not related to the normative system, hence, is neutral, i.e.:

$$w \vDash PA \text{ iff there is } u \in W, \text{ such that } Q(w,u) \text{ and either } u \vDash A \text{ or } A \notin R_u.$$

The above presented semantics constitute a particular combination of the possible-worlds semantics with the relating semantics. The semantics of the latter type were discussed in detail in [9], and its specific cases in [11]. The basis of such semantics is the evaluation of connection, i.e. the function defined for a given intentional functor α of arity n, mapping n-th Cartesian product of the sets of formulas of a given language into a set of elements representing the values of connection values between the given sentences:

$$f\alpha : \text{For} \rightarrow \text{VC},$$

where For is a set of formulas of a given language, and VC is a non-empty set of the connection values. In the case of deontic language, the matter involves two unary intensional functors – deontic operators. Consequently, in each world, we can introduce an evaluation of connection with two connection values. Such evaluations determine the subsets of formulas in each world, on the basis of the indicator function. In this particular case, the evaluation of connection becomes similar to the awareness function introduced by Fagin and Halpern [4] within the semantics of epistemic logic. Notice that the above-given truth-conditions of the deontic operators differ from the conditions introduced by Fagin and Halpern [4, p. 53] for the epistemic operator. Moreover, contrary to Fagin and Halpern, we introduce into the language neither the alethic modalities, nor any particular kind of operator which would constitute a linguistic equivalent of the new element of the model.

The work by Jarmużek and Klonowski [11] analyses models of relating deontic logic, in which instead of an indexed family of subsets of formulas, an indexed family of binary relations occurring between the formulas was considered. In this case, the unary approach was defined within the binary approach; that is, the family of subsets of formulas indexed by possible worlds was defined by means of the family of binary relations defined on the set of formulas indexed by possible worlds. Hence, the relation with the normative system was defined by relating the sentences. Such an approach becomes clear with regard to the analysis of deontic contexts through reference to various binary relations, such as: causal relation, time sequence, relations between action and sanction or action and issue of a relevant document, etc.

Needless to say, the binary relation, defined by the formulas, constitutes a special case of the evaluation of connection. The semantics based on such a relation constitutes a special case of the relating semantics obtained through limiting the evaluation of connection to the function defined on a Cartesian product of the set of the formulas with a bivalent codomain. Such a relational semantics probably has its origin in the work of Epstein [2]. An example of its application may be the analysis of the content relationships which is the foundation of the so-named relatedness logics and dependence logics defined by Epstein [2], [3, pp. 61-84, 115-143] with some particular conditions imposed on the models. A more general approach – where the starting point are models containing all binary relations specified on the set of formulas – proposed by Jarmużek and Kaczkowski [10] and explored by Jarmużek and Klonowski [12] (cf. [9]).

4. Jan Woleński's Metaethics

In the metaethics of Jan Woleński, the following two theses come to the fore:[3] (i) naturalism, and (ii) non-linguistic conception of norms. These theses are independent, and their combination is not common; however, the main idea is that the latter supports the former of more general nature.[4]

Jan Woleński's metaethical naturalism is notably a consequence of his broader argumentation for naturalism in philosophy (see [21]). However, Woleński also presents detailed metaethical arguments for naturalism, which at the same time solve his key issues, as well as arguments against antinaturalism. We will only briefly outline the most important line of argumentation in which occurs (ii).

In his metaethical works, Woleński devotes a lot of attention to the so-called Hume's guillotine, setting it, in a way, in the centre of metaethical considerations. Let us recall the well-known problem in the words of John Searle [16, p. 43]:

> It is often said that one cannot derive an "ought" from an "is". This thesis, which comes from a famous passage in Hume's *Treatise*, while not as clear as it might be, is at least clear in broad outline: there is a class of statements of fact which is logically distinct from a class of statements of value. No set of statements of fact by themselves entails any statement of value. Put in more contemporary terminology, no set of *descriptive* statements can entail an *evaluative* statement without the addition of at least one evaluative premise. To believe otherwise is to commit what has been called the naturalistic fallacy.

The last sentence explains the meaning of Hume's comments on the naturalistic metaethics. Woleński indicates that the problem can be generalised, and simply the relations between normative and descriptive sentences can be discussed. If the normative sentences will be understood as in the deontic logic, that is, with the operators "it is permitted that", "it is obligatory that", "it is indifferent that" and optionally with other ones, then the generalised Hume's thesis, according to Woleński, takes the following form (where sentence A is descriptive, non-tautological and non-deontic, i.e. does not include deontic operator, "D" is one of the deontic operators, and "$\nvDash$" expresses that a semantic consequence relation $\vDash$ doesn't hold[5]):

(1) $A \not\vdash DA$
(2) $DA \not\vdash A$.

According to Woleński, the generalised Hume's thesis "can be described as a thesis of logical separation of being (facts) and obligation" [20, p. 33]. Both constituents of the thesis should be considered in the naturalistic metaethics. Note that both the SDL and the relating deontic logic do satisfy (1) and (2), provided that they are expressed in the object language.[6]

He bases his deliberations on two axes of dispute in metaethics (see e.g. [19, p. 246]): naturalism vs antinaturalism and cognitivism vs noncognitivism. To put it simply, the naturalist believes that norms are part of the empirical reality, and the antinaturalist places them outside the empirical reality. The cognitivist assigns logical values to norms, and the noncognitivist believes that they have no logical values (various forms of irrealism). In defending naturalism, Woleński is not explicitly in favour of cognitivism or noncongnitivism, as he challenges their underlying assumption that norms are linguistic entities. Thus, in a way, he shifts the issue of truth and falsehood from norms – as in the dispute between cognitivism and noncognitivism – to normative sentences.[7]

The thesis about the non-linguistic character of norms is crucial in Woleński's argumentation. It presents four negative arguments in its favour, i.e. stating what norms are not – they are not linguistic entities; and one positive argument, i.e. stating what norms are (see, e.g. [20, p. 39]). The first three refer to linguistic practice (especially legal practice) and point to a categorical error: when we say that we comply with norms, that a norm applies, or that norms have social causes and effects, we do not mean linguistic expressions, we do not refer to sentences (cf. [15, p. 26]). The fourth argument is grammatical: we distinguish declarative, interrogative, and imperative sentences rather than normative sentences, which means that the latter must be reduced to one of these three types. Woleński argues that the choice of two types: declarative sentences (cognitivism) and imperative sentences (noncognitivism) results in problems for these standpoints.[8]

The positive argument indicates what standards are if they are not of a linguistic character. Woleński's idea, also developed in his works with Kazimierz Opałka, involves extending Austin's concept of performatives to the normative sphere. In short: "We claim that normalisation is an act of some kind, a norm is the result of such an act, and a normative utterance – the expression of a norm" (see [15, p. 27]).

Consequently, according to Twardowski's division into acts and their products, there are three components: the act of normalisation, the product of the act in the form of a norm, and normative utterance related to the norm (the expression of the norm). The naturalistic consequences are easy to identify: norms are not from a non-empirical reality, but are the products of the decisions taken by the norm-maker and the performative acts related to them, that is, certain actions in the world. Every norm was once established by someone (also collectively understood social entities) through a performative act. This approach is not burdened by the categorical error mentioned above: when referring to the validity, observance or application of a norm, we refer to the corresponding relation to the normative product of the performative act.

Although norms are not linguistic expressions, they can be communicated by means of linguistic expressions. Such utterances take the following general form:

(*) I order (prohibit, permit in terms of making it indifferent) A.

As we know, Austin did not attribute logical values to performative utterances; instead, he referred to the conditions of their effectiveness: they are effective if a number of factual and formal conditions is met. Woleński solves this problem by distinguishing the performative, that is, a certain action, from a performative utterance. "Effectiveness is not a matter of statements, but of actions. Provided that a given performative is effective, the relevant performative utterance is true, e.g. the sentence *I order that A* is true if effectiveness conditions for effectiveness of obligations are met" [20, p. 41].

The performative utterances that fall within (*), are called "primary normative utterances". A set of such statements together with their logical consequences – plus possible restrictions, such as non-contradiction – form a normative system. On the other hand, deontic sentences are "secondary normative utterances", and their logical value depends on the logical value of the primary normative utterances. "A normative system can also be defined as a set of true deontic sentences and their logical consequences, relativised to the given normalisation" [20, footnote 33].

Since both primary and secondary normative utterances constitute declarative sentences, ergo, bear logical value – when certain additional standard conditions are met, e.g. elimination of indexicality – there is no need to introduce the norm as a new semantic category. Since all the components of such a theory are elements of empirical reality, the result is a naturalistic stance.[9]

The combination of naturalism with the non-linguistic concept of norms results in a coherent metaethical stance, which Woleński combines with the classical approach to the deontic sentences expressed in SDL. Such a combination is not necessary, but constitutes a certain methodological requirement respected by Woleński on many other occasions: a philosophical stance should be consistent with the basic logical representation of given concepts, e.g. based on the generalised square of opposition or related to correctly interpreted modal (most often normal) logics. Compliance with such a requirement is an important advantage of Woleński's philosophy (including metaethics). However, it should be remembered that such basic logic faces many issues – shown above on the example of SDL in section 2. While solving these problems, relating deontic logic retains selected logical values of deontic concepts.

5. Normative Inferences, Metaethical Naturalism and Deontic Relationship

One of the fundamental metaethical issues is the problem of the validity of normative inferences. Having defined the basic normative concepts, we would like to employ them in conducting inferences. However, according to non-cognitivists, norms do not carry logical values; thus, they cannot be directly implemented into inferences. We have seen that in metaethics this problem is seen as associated with Hume's scepticism, whereas within the field of deontic logic, it appears from the beginning in the form of the so-called Jörgensen's Dilemma. Let us recall it in its original form [13, p. 290]:

> So we have the following puzzle: According to a generally accepted definition of logical inference only sentences which are capable of being true or false can function as premises or conclusions in an inference; nevertheless it seems evident that a conclusion in the imperative mood may be drawn from two premises one of which or both of which are in the imperative mood. How is this puzzle to be dealt with?

Let us note that Hume's guillotine is usually limited – as happened in Searle's words quoted above – to the situations when among the premises, there is not at least one normative premise. From the perspective of logic, however, it does not have much meaning: if one, non-exclusive premise and conclusion do not carry logical values, it is not possible to evaluate the validity of the reasoning. Hence, Jörgensen's approach is more general: it considers possible inferences, while Hume's approach was an expression of scepticism towards the theory of morality as such.

The fact is that we perform inferences, in which normative sentences play the main or indirect role:

(1) While driving his car, John turned right.
(2) There was an obligation to turn left there.[10]
(3) John broke the traffic laws.

There is no doubt that the sentence (1) can be assigned a logical value (provided that we interpret properly indexicality, vagueness etc.) – it is a sentence about a certain event occurring in the world. The sentence (2), which describes the traffic rule, raises more doubts. Of course, there is probably an appropriate road sign in the place referred to in this sentence, but this sentence does not simply speak of its presence in this place but states the existence of a corresponding norm. Also, the sentence (3) is not merely a sentence about an event in the world, but refers to the connection of such an event with the norm expressed in the sentence (2) –

similar conclusions can also be drawn, e.g. "John should not turn right". So, can we assign logical values to sentences (2) and (3)?[11]

The existence of such inferences in legal or everyday contexts, constitutes an indirect but quite strong argument in favour of the fact that these sentences bear logical values or, in a way, are related to the sentences that bear logical values. Such an assumption is also made within relating deontic logics, with a remark that such inferences are limited to the given normative system and are performed only within its boundaries. In other words, sentences should be related to the same normative system.[12] The advantage of such an approach is that it allows us to avoid SDL problems.

Consequently, does the normative system constitute a set of only the sentences that bear logical values? If so, then how is this set determined, namely: what constitutes a deontic relationship within this set? If not, then what else can constitute the elements of this set? Within logic, it is not necessary to determine this, and it is its unquestionable strength. However, in order to build a complete metaethical theory, at the same time, we have to look for an answer to solve Jörgensen's Dilemma.

Woleński, as mentioned in the previous section, understands the normative system as a set of true normative sentences (primary or secondary) limited to a given normalisation. Nevertheless, it is worth stressing that all the circumstances related to the conditions of effectivity of the relevant normative performatives are important for the constitution of such a set. Declarative sentences describe these circumstances, e.g. if I order someone to turn right, then one of the conditions is that this turn was permitted, that is, for the following sentence to be true "On such and such a road, and in such and such a place there is a right turn." These sentences are not part of the normative system but are related to the normative system. It is easy to notice the application of such an approach on the grounds of relating deontic logic, where it is assumed in the interpretation of deontic operators that the constituent sentences are related to the normative system. Thus – in order to preserve the basic features of Woleński's naturalistic metaethics – we should not understand it narrowly, as belonging to the system, but broadly, as being in relation to the normative system. In relation to the constitution of the effectivity of performative acts, which constitute truth-conditions of the primary normative sentences, which, in turn, are truth-conditions for deontic sentences.

Such an approach provides intuitive criteria of the validity of the normative reasonings. Firstly, if these sentences consist of deontic phrases or somehow depend on the validity of the norms, then they have logical values that depend on the effectivity of normative performative acts of the norm-maker.[13] Secondly, the sentences used in the inference should be related to the normative system. Consequently, in fact, most of the common normative inferences have an enthymematic character. In the example considered above, these are the sentences that lead to the effectivity of the performative act of the manager of a given road, that is, e.g. it had a legal foundation, but also the factual circumstances, that is, e.g. that actually there was a turn and a road, etc. A moment of reflection is enough to consider such consequences as natural and really related to the normative inferences.[14]

6. Summary

The value of philosophical logics lies mainly in the fact that they can constitute a common ground for philosophical dispute, providing tools to describe the aporias occurring there. Nevertheless, it happens that such logics exclude certain stances, indicating their contradiction or undesirable consequences.

In Jan Woleński's philosophy, it is essential that the proposed solutions are consistent with the basic logical properties of the analysed concepts. In the metaethical approach, Woleński emphasises the relations from the generalised square of opposition and Hume's principle. These are the minimum requirements that lead to standard deontic logic – naturalism is, thus, logically consistent. It is well known, however, that such a simple logic faces many problems that would also affect the given metaethical naturalism. The relating deontical logics described herein allow us to address specific problems, and at the same time, they acquire a philosophical interpretation related to naturalism justified by the non-linguistic concept of norms, which allows us to respond to Jörgensen's Dilemma and work out its informal details.

References

1. Carmo, J., and A. Jones. Deontic logic and contrary-to-duties, In D. M. Gabbay and F. Guenthner (eds.), *Handbook of Philosophical Logic*, vol. 8, Springer Science+Business Media, 2002, pp. 147-264.
2. Epstein, R. Relatedness and implication, *Philosophical Studies* 36 (2), 1979, pp. 137-173.
3. Epstein, R. (with the assistance and collaboration of: W. Carnielli, I. D'Ottaviano, S. Krajewski, R. Maddux). *The Semantic Foundations of Logic. Volume 1. Propositional Logics*, Dordrecht: Springer Science+Business Media, 1990.
4. Fagin, R., and J. Y. Halpern. Belief, awareness, and limited reasoning, *Artificial Intelligence* 34, 1988, pp. 39-76.
5. Hilpinen, R. Deontic logic, In L. Goble (ed.), *Blackwell guide to philosophical logic*, Oxford: Blackwell Publisher Ltd., 2001, pp. 159-182.
6. Hilpinen, R., and P. McNamara. Deontic logic. A historical survey and introduction, In D. Gabbay, J. Horty, X. Parent, R. van der Meyden and L. van der Torre (eds.), *Handbook of Deontic Logic and Normative Systems*, London: College Publications, 2013, pp. 3-136.
7. Hintikka, J. Deontic logic and its philosophical morals, In J. Hintikka, *Models for Modalities*, Dordrecht: Springer, 1969, pp. 184-214.
8. Jadacki, J. J. The Polish 20th century philosophers' contribution to the theory of imperatives and norms, *European Journal of Analytic Philosophy* 7 (2), 2011, pp. 106-145.
9. Jarmużek, T. Relating semantics as fine-grained semantics for intensional propositional logics, In A. Giordani and J. Malinowski (eds.), *Logic in High Definition. Current Issues in Logical Semantics*, London: Springer 2020, pp. 19-36
10. Jarmużek, T., and B. Kaczkowski, On some logic with a relation imposed on formulae. Tableau system F, *Bulletin of the Section of Logic* 43 (1/2), 2014, pp. 53-72.
11. Jarmużek, T., and M. Klonowski, On logic of strictly-deontic modalities. A semantic and tableau Approach, *Logic and Logical Philosophy* 29 (3), 2020, pp. 335-380.
12. Jarmużek, T., and M. Klonowski. Some intensional logics defined by relating semantics and tableau systems, In A. Giordani and J. Malinowski

(eds.), *Logic in High Definition. Current Issues in Logical Semantics*, London: Springer, 2020, pp. 37-54.
13. Jörgensen, J. Imperatives and logic, *Erkenntnis* 7, 1937, pp. 288-296.
14. Opałek, K. Argumenty za nielingwistyczną koncepcją normy. Uwagi dyskusyjne, *Studia Prawnicze* 3–4, 1985, pp. 205-218.
15. Opałek, K., and J. Woleński. Logika i interpretacja powinności, *Krakowskie Studia Prawnicze* 21, 1988, pp. 13-29.
16. Searle, J. How to derive *ought* from *is*, *Philosophical Review* 73 (1), 1964, pp. 43-58.
17. Woleński, J. Spór o znaczenie normatywne, In *Naturalistyczne i antynaturalistyczne interpretacje humanistyki*, Poznań: Wydawnictwo UAM, 1966, pp. 3-14.
18. Woleński, J. *Z zagadnień analitycznej filozofii prawa*, Warszawa, Kraków: PWN, 1980 (Wydanie drugie, zmienione i rozszerzone: Kraków, Wydawnictwo Aureus, 2012).
19. Woleński, J. Naturalizm, antynaturalizm i metaetyka, *Folia Philosophica* 29, 2011, pp. 241-256.
20. Woleński, J. Próba naturalistycznej interpretacji norm i ocen, *Etyka* 50, 2015, pp. 29-47.
21. Woleński, J. *Wykłady o naturalizmie*, Toruń: Wydawnictwo Naukowe Uniwersytetu Mikołaja Kopernika w Toruniu, 2016.

Notes

1. Clearly, both approaches retain their independence – taking one of them does not force adoption of the other – however, combining them into a uniform framework, although merely outlined herein, provides a new tool for analysing normative reasoning.
2. More on this subject in [11, section 2.2].
3. Metaethical issues were the subject of Jan Woleński's work from the beginning of his scientific career (see [17]), summarised by the book *Z zagadnień analitycznej filozofii prawa* (see [18]; new, revised and extended edition: Woleński 2012). Some ideas, mainly the non-linguistic conception of norms, he developed in collaboration with Kazimierz Opałek (see [15]). Opałek [14] also defended it independently. The short

description below we have based mainly on the newest publications: [19], [20], [21]. Woleński repeatedly points out that his defence of naturalism in metaethics is not categorical, thus – in other words – it is mostly a consequence of some set of abductive arguments. The contribution of Polish philosophers, including Woleński, to metaethics is discussed in a review work of Jadacki [8].

4. Woleński also discusses metaethical issues related to the bonitive sentences, adopting the standpoint of axiological presentationism. In this study, in establishing the relations with deontic logic, we limit ourselves merely to describing the metaethics of normative sentences.

5. Woleński employs "deductibility symbol": $\vdash$, that is, a symbol of syntactic consequence. However, in the context of Hume's guillotine and related problems, we prefer to employ semantic consequence, since it assumes that the sentences carry logical values, while, on the extralogical basis, it is possible to imagine that something is syntactically deductible (by simply performing acceptable transformations of the original schemes), and at the same time is neither true, nor false.

6. That is, by means of material implication. If, in turn, we allow A to be tautological, then, in result, we get one of the SDL problems which can be easily eliminated within relating deontic logic, by replacing the material implication with the relating implication, in which truth depends on the logical value of the constituent sentences and the occurrence of the relations between them (see [10], [12]).

7. "My preferences rather lie with noncognitivism, mainly because, nonetheless, the settlement of ethical disputes differs from the settlement of empirical disputes. On the other hand, as mentioned before, there seem to be no rational reasons to deny the axiological sentences the value of truth or false. However, this must be done with full understanding that it is not a matter of correlation between these sentences and natural reality in a narrow sense, but of truth in an appropriate deontic model relative to performatives, or in a bonitive model relative to axiological presentations" [20, p. 46].

8. Can't we also consider the interrogative sentences? In its direct form this would probably be challenging, but it is not out of the question that

standards can be related to a set of answers to a certain question. This concept is not further discussed herein.

9. And yet, does the non-linguistic concept of norms actually somehow force metaethical naturalism? While certain doubts arise at this point, it is worth noticing that even though the norm as an act constitutes a component of empirical reality, one of the conditions for the effectivity of such an act may be from outside of such reality. In other words, it is not impossible for the normative performative act to be a kind of transfer of the norm from a non-empirical into empirical reality, to be in a way "embodiment of a (proper) norm". Additional arguments are needed to weaken the occurrence of such a possibility. So, it may not be as easy to give up transcendence as imputed to naturalists.

10. This sentence can be formulated in a similar or equivalent way (e.g. whether we formulate a normative rule in general, or as one concerning John etc., nuances are not relevant here): John should have turned left; Left turn was obligatory; John was required to turn left, etc.

11. The above reservations are also made in favour of attempts to formalise a broad category of axiological sentences, including bonitive, evaluative, as well as imperative sentences and directives. The above example can be accordingly modified.

12. Needless to say, these sentences, individually, may also belong to other normative systems, but then – which is very intuitive – the validity of the inference cannot be considered. Are inter-normative inferences allowed, i.e. when the components of the inference belong to different normative systems? Perhaps, as far as they at least intersect.

13. Referring to our exemplary inference: what performative acts are behind the truth of sentence (2)? It is a performative act performed by the road manager, who, on the basis of the result of other performative – here: legislative (legal act, regulation) – established traffic rules in the described location.

14. Of course, some of them are shared with other common inferences, which are usually simplified, concealing the premises that are clear for interlocutors, not announcing the conclusion. etc. Thus, the characteristic attribute of normative inferences are psychological conditions, which indicate, for example, that the norm-maker actually has an intention to

create such and not another law, that he has appropriate powers to do so, etc.

A Note on Intended and Standard Models

Jerzy Pogonowski

Adam Mickiewicz University in Poznań
Szamarzewskiego 89/AB Street
60-568 Poznań, Poland
e-mail: pogon@amu.edu.pl

1. The Distinction: Intended Model versus Standard Model

Mathematical theories may concern either a specified structure or a class of structures. Examples of theories of the first kind include theories of fundamental number systems (natural numbers, integers, rational numbers, real numbers, complex numbers), certain systems of geometry (for instance Euclidean geometry), and possibly also set theory, at least at the early stage of its development. Theories of the second kind include theory of groups, fields, topological spaces, vector spaces, and so on. The distinction in question applies to modern mathematics, it does not make sense in the case of mathematics before the second half of the 19th century.

The notions of intended, standard and non-standard models may be applied in the case of theories of the first kind, for obvious reasons. The terms 'intended model' and 'standard model' are used sometimes interchangeably in literature. I propose to distinguish them in the following manner. The intended model of a theory is a structure which motivated the development of the theory in question. As a rule, this structure has been investigated for a long time and its properties are based on well-established mathematical intuitions emerging from the research practice.

A necessary condition for a structure to become an intended model is thus its domestication in the mathematical research. One could also say that intended models are cognitively accessible to a high degree. Then there emerges a theory of such a structure, ultimately an axiomatic theory.

The above characterization of the concept 'intended model' is intuitive, which in turn implies that the concept itself is also intuitive. A prominent example of an intended model in this sense is the natural number series with arithmetical operations defined in the usual way. Rational, real and complex numbers (as understood before the construction of the corresponding axiomatic theories of such numbers) provide further examples. It seems that the universe of the naive set theory could also be considered an example in this respect.

The notion of a standard model, in turn, may be introduced only after the theory in question has become a fully formalized theory, with overtly specified primitive terms and axioms characterizing them. In this situation the class of all models of the theory in question can be established. This class may consist of only one model or of many models, which depends on the language of the theory and the underlying logic, among other aspects. In the first case we obtain the standard model at once. In the second case we may only choose one of the models and call it standard. I propose to call a model 'standard', if it is most closely related to the intended model. The similarity between intended and standard model should be based on a kind of isomorphism. Because the standard model of a theory is a specific element of the well-defined class of all models of the theory in question, it is a genuine mathematical object and as such it is well-defined, too. We should remember, however, that the name *standard* was given to it on the basis of our decision. The latter was supported by the observed resemblance of the standard model to the intended model given in advance. It may also happen that certain theorems concerning the standard model provide additional support for our decision. Still, the selection of the name *standard* is based primarily on pragmatic criteria.

The standard model of arithmetic is determined uniquely (up to isomorphism) on the basis of second-order Peano axioms. In the case of first-order Peano arithmetic its standard model is only one of the continuum many countable models of this theory. According to Tennenbaum's theorem, it is the only recursive model of this first-order theory. It is also its prime model, meaning that it can be elementarily embedded in any other model of the theory in question. Non-standard models of arithmetic contain infinitely large numbers.

The completely ordered real field (satisfying thus the upper bound property) is determined uniquely (up to isomorphism). It is commonly accepted as the standard model of the arithmetical continuum. It is also a maximal Archimedean field but it is not algebraically closed. The complex field, in turn, is determined uniquely (up to isomorphism) as the only algebraically closed field of the characteristic zero whose transcendence degree over the field of rational numbers equals the continuum. No order compatible with the arithmetical operations is possible in the field of complex numbers.

The (first-order) theory of real closed fields is semantically complete, meaning that all models of this theory are elementarily equivalent, i.e. have the same set of true sentences. The real numbers, which form a real closed field, are thus characterized uniquely with respect to elementary equivalence in the first-order language.

The hyperreal field is also elementarily equivalent with the field of real numbers, but it is not an Archimedean field (it contains infinitesimals). The rather unfortunate name *non-standard analysis* given to the theory concerning the hyperreal field may suggest that hyperreal numbers are non-standard. However, it is mainly the matter of mathematical research practice to decide, on the basis of accumulated knowledge and fruitfulness of applications, which structure should be called standard.

A paper by Solomon Feferman [8] discusses the question of which formal representations of the geometric continuum could be thought of as standard. Feferman lists a few candidates: Euclid's continuum; Cantor's continuum; Dedekind's continuum; Hilbert's continuum; the continuum as the set of all branches in the full binary tree; and the continuum as the family *P(N)* (the full powerset of the set of all natural numbers). Feferman summarizes his paper on conceptions of the continuum as follows:

> Of all the conceptions of the continuum considered here, only those of sec. 3 stand as structural ones, and of those only 2^N and *P(N)* stand as *basic* structural conceptions. For, the continuum in Euclidean and Hilbertian geometry is not an isolated notion, while the continuum as given by Cantor's and Dedekind's construction of the real numbers, are hybrid constructions. The set 2^N of all sequences of 0s and 1s isolates the set-theoretical component of Cantor's construction, while the set *P(N)* of all subsets of *N* isolates

that of Dedekind's construction, but both of these lose entirely the basic geometric intuition of the continuum. On the other hand, it does not count against Cantor's and Dedekind's conceptions of the continuum in the form of the real number system R that they are hybrids of geometrical, arithmetical and set-theoretic notions. On the contrary, by a kind of miracle of synergy, R has proved to serve together with the natural numbers N as one of the two core structures of mathematics; together they are the *sine qua non* of our subject, both pure and applied.

If first-order Zermelo-Fraenkel set theory is consistent (which cannot be proved in the theory itself), then it has a plentitude of models. It is commonly accepted in the mathematical community to call a model of this theory *standard*, if the interpretation of the membership predicate in it is the real membership relation. Models of set theory without the axiom of foundation are usually seen as non-standard models.

The distinction between *genuine* (*normal, natural,* etc.) mathematical objects and those called *unintended* (*unwilling, imaginary,* etc.) was noticed in the history of mathematics even before the second half of the 19th century. For example, negative or imaginary numbers were long rejected as legitimate mathematical objects before they finally became accepted by the mathematical community. It is important to make a distinction between a *non-standard* (object) and an *innovation*. Haim Gaifman discussed the following *innovations* in mathematics in his paper [11] devoted to the non-standard models: the discovery of irrationals; the incorporation of negative and complex numbers in the numeral system; the extension of the concept of *function* in the nineteenth century; and the discovery of non-Euclidean geometry. Gaifman gives arguments that such innovations should not be considered non-standard. He also discusses certain further candidates for being a standard mathematical object, including *well-ordered* and *constructible* sets. The full powerset operation, on the other hand, escapes from the list of standards.

There are several ways of constructing non-standard models of mathematical theories. Let us consider Peano arithmetic (PA). If we expand its language by a new individual constant c and take into account an infinite set of sentences $C = \{\neg \bar{n} = c : n \in N\}$ (where $\bar{n}$ is the numeral denoting the natural number n), then each of its finite subsets has a model and it follows from the compactness theorem that

C itself has a model. The denotation of c in this model is different from each standard natural number and hence the model in question is non-standard. Another possibility, already anticipated by Thoralf Skolem, is to build a suitable ultraproduct (actually, an ultrapower) starting with the standard model of PA. One can also consider a full binary tree of expansions of arithmetic and show that each branch of this tree corresponds to a model of PA; one of them is the standard model, while all others are non-standard models. We will come back to the latter possibility below, discussing Jan Woleński's views on non-standard models.

2. On the Origin of Metalogical Concepts

Claims about uniqueness of models require precise tools of comparison of the models themselves. There are essentially two ways of characterizing the indistinguishability of models of a given theory. One of them is structural: we may ask whether the models are isomorphic (or partially isomorphic, or one of them being a homomorphic image of the other, and so on). The notion of isomorphism emerged in algebraic considerations in the early 19th century. Isomorphic structures are structurally indistinguishable. If all models of a theory T are isomorphic, then we say that T is a *categorical* theory. A theory T is *categorical in power* κ (where κ is an infinite cardinal number), if it has a model of power κ and all its models of power κ are isomorphic. It should be stressed that first-order theories cannot be categorical, with the exception of certain trivial cases. This is a consequence of Löwenheim-Skolem-Tarski theorem which says that if a theory (without finite models) has a model, then it has models of all infinite cardinalities.

Another kind of indistinguishability of models is based on semantic criteria. We say that two models are *elementarily equivalent*, if the sets of sentences true in them coincide. A theory T is (semantically) *complete*, if all its models are elementarily equivalent. If two models are isomorphic, then they are also elementarily equivalent, and hence categoricity implies semantic completeness, but the converse implication does not hold.

The notion of categoricity originated in the papers of Edward Huntington and Oswald Veblen. Huntington used the term *sufficiency* in 1902 and Veblen replaced it by the term *categoricity* in 1904. In the nineteen-twenties Abraham Fraenkel and Rudolf Carnap used the term *monomorphy* (*Monomorphie* in German) in the meaning in question. Fraenkel and Carnap considered also a kind of semantic completeness (called by Carnap *non-forkability*, in German: *nicht-Gabelbarkeit*). It should be stressed that before emergence of well-developed metalogic the notions of categoricity and semantic completeness were not sharply separated. In the absence of precise formal logical tools the claim that isomorphism implies semantic indistinguishability was understood evident by Huntington, Veblen and also earlier by Richard Dedekind. An important early contribution to the relationships between these notions is the paper [15] written by Lindenbaum and Tarski. Tarski's paper [22] from 1940 (printed as appendix in [16]) elaborates further this issue. Tarski introduced the notion of elementary equivalence in the nineteen-fifties. Many important observations concerning the emergence and mutual relations between the notions in question are contained in [1], [6] and [7].

Categoricity, categoricity in power and semantic completeness were further characterized in full detail in classical and modern model theory. There is no need to report on these results here; an interested reader may consult for example [14] or [17]. Let us only add that the tools from model theory are sufficient for talking about several kinds of indistinguishability of models and the uniqueness of these models.

3. Extremal Axioms

The term 'extremal axiom' was introduced in the paper [4] written by Carnap and Bachmann. The authors tried to present a general form of these axioms using the logical framework of the theory of types. At the beginning of the paper they write (citing [5] which is the English translation of [4]):

> Some important axiom systems are so constructed that first a series of axioms is given, making certain statements about the basic concepts of the axiomatic theory, and then at the end an axiom of a special sort appears which apparently speaks about the foregoing axioms and not about the special concepts of the theory. The most famous axiom system of this sort is Hilbert's axiom system of Euclidean Geometry. It ends with the famous 'completeness axiom' which runs as follows [The footnote given here by the authors reads: D. Hilbert, *Grundlagen*

der Geometrie (Leipzig and Berlin). We take the Hilbert completeness axiom in the form it has in editions 2–6, not the 'linear formulation' of the 7th edition of 1930. – J.P.]:

'The elements (points, lines, planes) of geometry constitute a system of things which cannot be extended while maintaining simultaneously the cited axioms, i.e., it is not possible to add to this system of points, lines, and planes another system of things such that the system arising from this addition satisfies axioms AI-V1.'

Axioms of this sort, which ascribe to the objects of an axiomatic theory a maximal property – in that they assert that there is no more comprehensive system of things that satisfies a given series of axioms – we call a maximal axiom. The same axiomatic role as that of maximal axiom is played in other axiom systems by minimal axioms which ascribe a minimality property to the objects of the discipline. Maximal and minimal axioms we call collectively extremal axioms [5, pp. 68-69].

Besides Hilbert's axiom of completeness in geometry (which was an axiom of maximality) Carnap and Bachmann considered two axioms of minimality: the induction axiom in arithmetic and Fraenkel's axiom of restriction in set theory. The latter says, roughly speaking, that only these sets exist whose existence can be proved in set theory (and hence the universe of all sets should be as narrow as possible). Extremal axioms were considered by Carnap and Bachmann as expressing a kind of completeness of models and hence as candidates for conditions characterizing models in a unique way. The famous limitative theorems proved later in the 20th century showed the possibilities and restrictions in this respect.

Early Carnap's views on extremal axioms and metalogic are best described in several papers written by Georg Schiemer (see for instance [21]). My book [19] presents logical, mathematical and cognitive aspects of extremal axioms. In particular, I propose to extend the inventory of extremal axioms by taking into account Kurt Gödel's axiom of contructibility, John von Neumann's axiom of the limitation of size and Roman Suszko's axiom of canonicity (these are examples of restriction axioms in set theory, hence axioms of minimality) as well as axioms of the existence of large cardinals in set theory (which are axioms of maximality). I also mention an interesting example of a maximality axiom in algebra, namely a generalization of Dedekind's axiom of continuity proposed by Philip Ehrlich and used by him to prove categoricity results concerning certain non-Archimedean structures.

Hilbert's axiom of completeness in geometry presented in [13] was later replaced by the axiom of continuity for real numbers which resulted, among others, in the proof of categoricity of the system of Euclidean geometry (see for example [3]). Second-order axiom of induction in arithmetic is used in the proof that there exists exactly one (up to isomorphism) Peano algebra. On the other hand, first-order Peano arithmetic is far from being semantically complete (and hence also categorical).

It is interesting that mathematicians have changed their views on extremal axioms in set theory. The axioms of restriction were abandoned, which was most explicitly shown in [10]. Set theoreticians are recently eager to investigate several axioms of the existence of large cardinals which presuppose that the universe of all sets should be as large as possible. Kurt Gödel himself opted for this trend and Ernst Zermelo proposed to accept the existence of the whole transfinite hierarchy of strongly inaccessible numbers already in his second axiomatization of set theory presented in [26].

4. Jan Woleński on Intended and Standard Models

Jan Woleński devoted several works to metatheoretical analysis of formalized theories. In my opinion, most interesting are his proposals involving applications of concepts elaborated in metalogic to the analysis in question. It is justified to claim that Jan Woleński achieved perfection in this work. He may doubtlessly be considered the leading continuator of the famous Warsaw-Lviv school.

We shall analyze in brief Woleński's views on intended and standard models. Our main source is his book on epistemology [25]. Many Polish philosophers wrote on intended models (notably Marian Przełęcki, Adam Nowaczyk, Ryszard Wójcicki, and Adam Grobler) but their analysis was focused mainly on intended models of empirical theories. Jan Woleński's reflections, in turn, are devoted mainly to intended and standard models of mathematical theories which is also the main issue discussed in this note.

Jan Woleński influenced my own views on intended and standard models mainly with respect to the opinion that these models are distinguished not on purely syntactic or semantic criteria but rather by taking into account also certain pragmatic factors. There may be small differences between his understanding of the distinction between intended and standard models and the one presented at the beginning of this note, but they are negligible.

Woleński recalls the construction of the tree of extensions of first-order Peano arithmetic PA ([25], 256; [18], 161). Let T_0=PA and let ψ_0 be any undecidable statement in T_0. We put: T_{00} = PA + ψ_0 and T_{01} = PA + $\neg\psi_0$. For any finite 0–1 sequence σ let: $T_{\sigma 0} = T_\sigma + \psi_\sigma$ and $T_{\sigma 1} = T_\sigma + \neg\psi_\sigma$, where ψ_σ is any undecidable sentence of T_σ (for any T_σ there exists such an undecidable sentence). We obtain in this way the full binary tree of extensions of PA. This tree has continuum many branches. It follows from the compactness theorem that the union of theories from each branch is consistent (under the assumption of consistency of PA) and hence each such union has a model. Further, due to the downward Löwenheim-Skolem theorem each such union has a countable model. No two such models are elementarily equivalent which follows from the construction of the above tree. Consequently, no two such models are isomorphic.

Let ψ_0 be identical with *Con(PA)* (that is, the sentence expressing the fact that PA is consistent) and let ψ_α express the consistency of T_α. Then the model of the leftmost branch of the above tree is isomorphic to the standard model of PA. All other branches have countable non-standard models. Each sentence of the form $\neg Con(T_\alpha)$ has the Gödel number which is a non-standard natural number in the respective model. Let us note on the margins that PA is a *wild* theory: it has, in each infinite power κ, the maximum possible number of models, that is 2^κ (provided the consistency of PA, of course).

The standard countable model of PA can be distinguished out of the totality of countable models of this theory only using some metatheoretical results, as already mentioned above. However, Jan Woleński proposes a more deep and subtle analysis of this issue. We need some auxiliary tools to present his views here:

A theory T is *descriptively complete* (in short: *o-complete*) with respect to a sequence $(a_s)_{s \in S}$ of individual constants (where S is any index set), if for any formula $\varphi(x)$ of the language of T with one free variable x the following implication holds: if $\varphi(x/a_s)$ is a theorem of T for all $s \in S$, then also $\forall x \varphi(x)$ is a theorem of T. If the sequence of individual constants in question is countable, then we say that T is *ω-complete*.

A theory T is *constructive* with respect to a sequence of terms $(t_s)_{s \in S}$, if for any formula $\varphi(x)$ of the language of T with one free variable x the following implication holds: if $\exists x \varphi(x)$ is a theorem of T, then $\varphi(x/t_s)$ is a theorem of T for some $s \in S$.

A theory T is *o-consistent* with respect to a sequence of terms $(t_s)_{s \in S}$, if for any formula $\varphi(x)$ of the language of T with one free variable x the following implication holds: if $\varphi(x/t_s)$ is a theorem of T for all $s \in S$, then $\exists x \neg \varphi(x)$ is not a theorem of T. If the sequence of terms in question is countable, then we say that T is *ω-consistent*. If a theory T is not ω-consistent, then we say that T is *ω-inconsistent*.

By the *ω-rule* we understand a rule of inference with an infinite set of premises $\varphi(\bar{0}), \varphi(\bar{1}), \varphi(\bar{2}), ...$ and the conclusion $\forall x \varphi(x)$.

These notions are related to the possibility of associating names with the elements of the domain of a model. ω-consistency was used already by Kurt Gödel in the formulation of his first incompleteness theorem. Descriptive completeness and constructivity were used by Andrzej Grzegorczyk in his famous paper on categoricity [12]. If the language of our theory contains numerals, then we can talk in this language about specific natural numbers. There arises a question of how these properties can be used in the characterization of models of a theory.

For any model M let *Th(M)* denote the *theory of M*, that is the set of all sentences true in M. Let N_0 denote the standard model of PA, N_c the non-standard model obtained by using the compactness argument in the way described above and N_{in} the non-standard model of the theory $PA + \neg Con(PA)$ obtained from the tree of expansions of PA presented earlier. The set *Th(N₀)* is thus the set of all arithmetical truths, that is true sentences about standard natural numbers. We recall that PA is incomplete and essentially undecidable. It is not finitely axiomatizable. If we add the infinitary ω-rule to PA, then the enriched theory becomes complete, but the price for that is very high, because we admit infinitary proofs, which is of course a debatable decision.

Jan Woleński uses an original generalization of the traditional square of oppositions for a formal representation of the logical dependencies between the notions of consistency, inconsistency, ω-consistency, and ω-inconsistency. It should be noted that these generalizations (see [24]) appeared to be a very productive and effective tool of logical analysis as shown by Woleński in his numerous articles on analytical philosophy. We are interested here mainly in possibilities of applying the notions in question to the characterization of intended and standard models.

All axioms and theorems of PA are true in the model N_{in}. However, the sentence $\neg Con(PA)$ is also true in N_{in}. The Gödel number of this sentence cannot be a standard natural number because otherwise PA would prove its own inconsistency, contrary to what was assumed. The sentence $\neg Con(PA)$ is obviously false in the standard model N_0 and Woleński writes that it is difficult to express its sense in the language appropriate for talking about N_0. If we are looking for formal criteria of being the standard model of arithmetic, then a good candidate could be the well-ordering property of the set of natural numbers. Woleński shares this opinion with Haim Gaifman (see [11]).

The set $Th(N_0)$ of all standard arithmetical truths is ω-consistent, ω-complete and constructive with respect to the sequence of all numerals. Woleński argues that o-consistency and constructivity are too strong conditions for the characterization of an arbitrary set of true sentences. For example, the set $Th(N_{in})$ is consistent but ω-inconsistent. It cannot be constructive, because consistency and constructivity imply ω-consistency. Further, Woleński adds that it is possible to consider the set $Th(N_c)$ as o-consistent and constructive with respect to a suitably chosen sequence of constants. Then $Th(N_c)$ is also o-complete. Woleński concludes from this that consistency (even maximal consistency) and o-completeness are minimal syntactic conditions characterizing the set of sentences true in any model and that the existence of theories which are consistent but at the same time ω-inconsistent clearly shows that truth differs essentially from provability. The semantic theory of truth alone is unable to distinguish the standard model in the class of all models.

Woleński says a few words explicating the commonly accepted assumption that PA is (a formal representation) of the True Arithmetic. From the point of view of a mathematician this could mean that the True Arithmetic is simply the totality of all logical consequences of the axioms of PA, even if not all of them have real applications. Another position (taken by a logician, according to Woleński) could accept the set $Th(N_0)$ as the True Arithmetic, thus identifying it with all arithmetical truths. Non-standard models of arithmetic can nevertheless be fundamental in certain mathematical disciplines – a notable example is the hyperreal field which has become recently more and more important in mathematical analysis.

Woleński expresses a few interesting remarks concerning the ways of formalization of arithmetic. The class of models isomorphic to N_0 can be characterized in second-order logic and this fact is considered a virtue of such formalization, first of all by the professional mathematicians. However, second-order arithmetic is undecidable and incomplete. The great expressive power of second-order logic is related to the acceptance of the absolute notion of a set. The expressive power of a logic is inversely proportional to its deductive power. Jan Woleński explicitly opts for first-order formalization, which possesses a lot of 'good' deductive properties and adds that this choice does not have any influence on the criteria of standardness of models.

The monograph [25] contains a very detailed analysis of the notion of an *analytic sentence*. One type of such sentences is relevant to standard models. Woleński proposes to call a sentence ψ *analytical in the pragmatic sense*, if there exists a theory T such that ψ is a theorem of T and ψ is true in the intended model of T. From the formal (logical) point of view standard models are as good as non-standard ones. It is our epistemic decision to call a model standard. We have argued in the first part of this note that this decision is determined by reflecting on the properties of the intended model, a structure investigated prior to the emergence of the formal (axiomatic) theory.

The monograph in question contains also a critique of Putnam's arguments expressed in [20]. Jan Woleński shows that Putnam is wrong claiming that models are nothing else but constructions inside theories. Putnam assumes that we refer to models (in particular to the intended model) always using the tools of the corresponding theory. This is clearly false, writes Woleński, because we must refer to metatheory when distinguishing between models. This is obvious for instance in the explication of Skolem's paradox in the context of models of the theory of real numbers. We switch to metatheory asserting that the proper (adequate) model of this theory has a power of continuum. The impossibility of definition of models in the object language, which follows from metalogical results, is discussed in more detail in [23].

5. Concluding Remarks

The main goal of this note was to present Jan Woleński's views on intended and standard models of mathematical theories. His contribution to this issue is based on an original application of metalogical results to philosophical problems. One can hardly find in philosophical literature examples of formal analysis comparable in depth and subtlety to those provided by Jan Woleński. My own distinction between intended and standard models was influenced by his proposals. In a sense, the distinction in question slightly resembles the distinction between the intuitive notion of a computable function and any precise mathematical representation of computability (for instance recursive functions or Turing machines).

Woleński's remarks are related first of all to models of arithmetic and to a lesser extent to geometric continuum and set theory. Taking into account the history of mathematics on a large timeline it seems legitimate to say that the intended model of arithmetic is much better understood than the continuum. The long philosophical debate about the structure of a continuum is still vivid and far from ultimate conclusions. The most commonly accepted representation of the geometric continuum by the arithmetical continuum of real numbers competes with the quite new representation based on hyperreal numbers. One can also find the opinion that the continuum should not be considered as a set of points, though no well-developed mathematically correct alternative is in sight at the present moment. This situation may prompt us to the conclusion that mathematicians have described several aspects of the continuum but have not captured ***the*** intended model of the continuum yet. A very interesting recent review of opinions on the structure of the continuum can be found in [2]. The discussion concerning models of set theory is also far from being closed as is clearly visible from the research directed towards new axioms which could characterize the set-theoretical universe in a more unique way.

References

1. Awodey, S., and E. H. Reck. Completeness and categoricity. Part I: Nineteenth-century axiomatics to twentieth-century metalogic, *History and Philosophy of Logic* 23, 2002, pp. 1-30.

2. Bedürftig, T., and R. Murawski. *Philosophy of mathematics*, Berlin Boston: Walter de Gruyter GmbH, 2018.
3. Borsuk, K., and W. Szmielew. *Podstawy geometrii*, Warszawa: Państwowe Wydawnictwo Naukowe, 1975.
4. Carnap, R., and F. Bachmann. Über Extremalaxiome, *Erkenntnis* 6, 1936, pp. 166-188.
5. Carnap, R., and F. Bachmann. On extremal axioms [English translation of Carnap and Bachmann 1936, by H.G. Bohnert], *History and Philosophy of Logic* 2, 1981, pp. 67-85.
6. Corcoran, J. Categoricity, *History and Philosophy of Logic* 1, 1980, pp. 187-207.
7. Corcoran, J. From categoricity to completeness, *History and Philosophy of Logic* 2, 1981, pp. 113-119.
8. Feferman, S. Conceptions of the continuum, *Intellectica* 51, 2009, pp. 169-189.
9. Fraenkel, A. A. *Einleitung in die Mengenlehre*, Berlin: Verlag von Julius Springer, Berlin, 1928.
10. Fraenkel, A. A., Y. Bar-Hillel, and A. Levy. *Foundations of set theory*, Amsterdam London: North-Holland Publishing Company, 1973.
11. Gaifman, H. Nonstandard models in a broader perspective, In A. Enayat and R. Kossak (eds.), *Nonstandard models in arithmetic and set theory*. AMS Special Session Nonstandard Models of Arithmetic and Set Theory, January 15–16, 2003, Baltimore, Maryland. *Contemporary Mathematics* 361, Providence, Rhode Island: American Mathematical Society, 2004, pp. 1-22.
12. Grzegorczyk, A. On the concept of categoricity, *Studia Logica* 13, 1962, pp. 39-66.
13. Hilbert, D. *Grundlagen der Geometrie*, Festschrift zur Feier der Enthüllung des Gauss-Weber-Denkmals in Göttingen, Leipzig: Teubner, 1899.
14. Hodges, W. *Model theory*, Cambridge: Cambridge University Press, 1993.
15. Lindenbaum, A., and A. Tarski. Über die Beschränkheit der Ausdruckmittel deduktiver Theorien, *Ergebnisse eines mathematischen Kolloquiums 1934–1935*, 7, 1936, pp. 15-22.
16. Mancosu, P. *The adventure of reason. Interplay between philosophy and mathematical logic, 1900–1940*, Oxford: Oxford University Press, 2010.

17. Marker, D. *Model theory: an introduction*, New York Berlin Heidelberg: Springer-Verlag, 2002.
18. Murawski, R. *Funkcje rekurencyjne i elementy metamatematyki. Problemy zupełności, rozstrzygalności, twierdzenia Gödla*, Poznań: Wydawnictwo Naukowe UAM, 2000.
19. Pogonowski, J. *Extremal axioms. Logical, mathematical and cognitive aspects*, Poznań: Wydawnictwo Nauk Społecznych i Humanistycznych UAM, 2019.
20. Putnam, H. Models and reality, *The Journal of Symbolic Logic* 45, 1980, pp. 464-482.
21. Schiemer, G. Carnap on extremal axioms, 'completeness of models', and categoricity, *The Review of Symbolic Log*ic 5 (4), 2012, pp. 613-641.
22. Tarski, A. On the completeness and categoricity of deductive theories. Appendix in Mancosu, P. *The adventure of reason. Interplay between philosophy and mathematical logic, 1900–1940*, Oxford: Oxford University Press, 2010, 1940, pp. 485-492.
23. Woleński, J. *Metamatematyka a epistemologia*, Warszawa: Wydawnictwo Naukowe PWN, 1993.
24. Woleński, J. Kwadrat logiczny – uogólnienia, interpretacje, In J. Perzanowski and A. Pietruszczak (eds.), *Logika & filozofia logiczna*, Toruń: Wydawnictwo Uniwersytetu Mikołaja Kopernika, 2000, pp. 45-57.
25. Woleński, J. *Epistemologia. Poznanie, prawda, wiedza, realizm*, Warszawa: Wydawnictwo Naukowe PWN, 2005.
26. Zermelo, E. Über Grenzzahlen und Mengenbereiche: Neue Untersuchungen über die Grundlagen der Mengenlehre, *Fundamenta Mathematicae* 16, 1930, pp. 29-47.

About Some New Methods of Analytical Philosophy. Formalization, De-formalization and Topological Hermeneutics

Janusz Kaczmarek

University of Łódź
Lindleya 3/5 Street
90-131 Łódź, Poland
e-mail: janusz.kaczmarek@uni.lodz.pl

1. Methods, Procedures, Rules, Operations

In this paper I refer to the work of Jan Woleński entitled "Kierunki i metody filozofii analitycznej" (Directions and methods of analytical philosophy) and in particular to its second part entitled "Methods of analytical philosophy". It discusses some methods characteristic for the analytical practice of philosophy, namely methods of: a) logical constructions (Russell, descriptive theory), b) explication (Carnap), c) paraphrases (Ajdukiewicz), d) presuposition (Strawson, Hart) and e) paradigm-case argument (Urmson, Hart) [25]. Of course, Professor Woleński has taken up the subject of methods in philosophy many times (comp [26] and [27]).

I will not discuss the methods indicated above, as such a description has been made many times [4], [7], [21], [25], [29]. On the other hand, I want to focus on newer methods or procedures of analytical philosophy, i.e. logical hermeneutics and topological hermeneutics, and I will try to show that some of the procedures considered within the phenomenological method are important for the

analytical study of philosophical problems. Therefore I will present below:
(a) specialization and generalization operations,
(b) Husserl formalization and de-formalization operations,
c) my own proposal, which I called topological hermeneutics and which I see as a complement to Wolniewicz's logical hermeneutics.
These methods will be partly confronted with the method of explication, paraphrases and logical constructions.

1.1. Note on Method, Procedure and Operation – Ambiguity of these Terms and/or Concepts

In many works we find descriptions of particular philosophical methods. Let us ask ourselves: what is a philosophical method? The answer is not easy, because when we look, for example, at the proposals of phenomenologists, one talks about the phenomenological method or methods, but also points to some special techniques (procedures, operations) such as eidetic reduction, *epoche*, variation or formalisation. It is similar in the framework of analytical philosophy, where the analytical method is talked about (aimed – following Bocheński – at language, analysis and logic), but also indicates some specific procedures such as Carnap's explication or Russell's descriptions.

Therefore, I propose that the method should be understood, in a working way, as a set of procedures characteristic of a given philosophical direction. A method understood in this way is then a set of detailed procedures, which I propose to call also tools or operations. Thus, for example, a phenomenological method is a specific way of reasoning and conducting research, in which we use (tools, operations) eidetic reduction, parenthesizing, variation, formalization, de-formalization, specialization and generalization (perhaps not everything yet). In turn, in the analytical method, i.e. the one characteristic of the analytical philosophy of the 20th century, we will encounter such tools and operations as: application of some logic (e.g. classic, temporal Scott's logic, modal S5, etc.), axiomatization (cf. Wolniewicz's axioms for the lattice of the situations), development or use of the logical square, formal approach to definite descriptions and many others. Interestingly, both the phenomenological method and the analytical one can be characterized in a general way emphasizing their main "attitude". For example,

Bocheński characterizes the analytical philosophy itself through keywords: language, analysis, logic and objectivity. From this we can conclude that the analytical method is characterized by: a) a turn to language and analysis of language, b) analysis of language using methods of logic, c) an attempt at objective analysis of what is on the side of reality and what can be expressed linguistically. Similarly, we can formulate basic axioms (or keywords) of phenomenology. Let us propose, therefore, at least the following postulates: a) turning towards the investigation of things, b) extracting what is essential (i.e., connected with the essence of the investigated thing), c) capturing what appears to our "self" as unreduced and free from any theoretical assumptions.

The brief proposal presented here may seem unjustified, but let us note that we find a similar approach in the book Bocheński [4]. Bocheński justifies that in contemporary philosophy we meet four basic methods [4, p. 14]:
1. the phenomenological method,
2. the language analysis,
3. the deductive method,
4. the reductive method.

In turn, in the book itself, Bocheński discusses in the following chapters the methods that correspond to the above, but are called respectively: the phenomenological method, semiotic methods, the axiomatic method and reductive methods. We have here some minor inaccuracies, because in the end we can ask: do we have a reductive method or rather (different) reductive methods; is language analysis the same as semiotic methods, etc.? From the text we learn, however, that Bocheński leans towards talking about the method as a specific style of conducting research that is most often appropriate for a given philosophical trend, while the terms procedure or operation should be used for more detailed tools. For example, eidetic reduction or *epoche* are called by Bocheński procedures, although he also uses the name "rule" [4, pp. 18-19].

The above mentioned demands do not aspire to a final solution. I just want to point out that the above problems call for a reliable and methodological reflection on philosophical methods and their detailed procedures (which I allowed myself to call also tools or operations). Therefore, in Part II we will give relevant examples of both analytical and phenomenological work.

1.2. Validation/Justification of Philosophical Methods

When Ajdukiewicz was proposing his method of paraphrases, he noticed that it should be justified, validated. I think that the problem of justification concerns every method, including any other presented in this work or discussed by Woleński, Bocheński, Stegmueller. Let us therefore look at the problem of validation in Ajdukiewicz's view. In the article *On the Applicability of Pure Logic to Philosophical Problems* from 1934 he writes:

> The apparent use of logic in solving philosophical problems formulated in natural language does not consist, therefore, in the deduction from logical theorems by legitimate substitution of conclusions which contribute to the solution of those problems. The procedure which has all the appearances of such application in fact consists in the construction in a natural language of sentences whose structure is isomorphic with the structure of logical theorems, i.e. in paraphrasing logical sentences into sentences with variables ranging over different domains of substitution than logical variables. It is only from such paraphrases that one may derive by substitution consequences relevant to philosphical problems formulated in a natural language. There is no doubt that the construction of such a system of sentences is desirable, for it would constitute the logic of ordinary language. However, those sentences, as paraphrases of universal logical sentences, require a validation which the existing contemporary logic is unable to supply.
> They could be validated as analytic sentences through a meaning analysis of the expressions of ordinary language. In the search for this validation one might use the phenomenological method. Alternatively, they could be justified by elevating them to the rank of postulates which – disregarding the meanings expressions have in ordinary language – would fix those meanings arbitrarily. This second method is more promising, it seems, than the phenomenological one which should be tried nevertheless. One must not forget, however, that if the second of the two methods is used the expressions of the language may

acquire meanings different from those they had previously. Hence the same verbal formulations might not express the same problems. However, this need not necessarily be regrettable (p. 93, The Scientific World Perspective).

Ajdukiewicz, as we can see, points to two paths leading to the validity of sentences being paraphrases of generalised logical sentences. The first one is to consist in the meaning analysis of sentences-paraphrases and treating them as analytical sentences. Then – in his opinion – the phenomenological method could be helpful. The second would consist in treating these sentences as postulates. Ajdukiewicz does not explain in detail what the application of the phenomenological method is to consist in. We can only guess that Husserl's analyses of expressions, meanings, senses, sentences, judgments proposed in *Logical Investigations* should be used. On the other hand, treating sentences (paraphrases) as postulates results in the unambiguity of terms but at the same time introduces arbitrary meanings that do not have to coincide with the meanings of expressions occurring in philosophical problems.

It is interesting that Bocheński also mentions the need to authorise (validate) the method. Bocheński directly writes about the justification of the phenomenological method, the justification of language analysis and the justification of formalism. I conclude from this that each method, and perhaps also the individual procedures of a given method, must make sure to reflect on their justification. For example, according to Bocheński, justification for formalism can be found in a) possibilities (thanks to formalism) of going beyond what is intuitively obvious, b) clear separation and explanation of concepts, c) elimination of hidden assumptions, and finally d) possibilities of different interpretation of what is formal and universal [4, pp. 40-41].

2. Husserl and Analytical Tools

Husserl, the founder of phenomenology, develops and uses the phenomenological method in his studies. In the initial pages of his Ideas I, however, he draws attention to some detailed tools (operations, procedures, rules), which are used or should be used by an ontologist (because here, in paragraphs 7 – 17, it is not so much about phenomenology as it is, above all, about formal ontology and regional ontologies). These tools are: specialisation, generalisation, formalisation and de-formalisation. Let us look at them and show that they are also tools used by analytical philosophers. I personally use them when I conduct ontological research.

Phenomenology is for Husserl a field of analysis through which one prepares the ground for particular sciences and philosophical problems. These analyses are aimed at examining the essence of various objects and the pure form of the object in general. The ontologist does the same – let us underline this – as well. Husserl writes about this subject in this way [8, p. 19] of the original edition:

> Any concrete empirical objectivity finds its place within a *highest* material genus, a "region," of empirical objects. To the pure regional essence, then, there corresponds a *regional eidetic science* or, as we can also say, ,a regional ontology. In this connection we assume that the regional essence, or the different genera composing it, are the basis for such abundant and highly ramified cognitions that, with respect to their systematic explication, it is indeed worth speaking of a science or of a whole complex of ontological disciplines corresponding to the single generic components of the region.

And then on [8, p. 19]:

> *Any science of matters of fact* (any experiential science) *has essential theoretical foundations in eidetic ontologies.* For (in case the assumption made is correct) it is quite obvious that the abundant stock of cognitions relating in a pure, an *unconditionally* valid manner to all possible objects of the region – in so far as these cognitions belong partly to the empty form of any objectivity whatever and partly to the regional Eidos which, as it where, exhibits a *necessary material form* of all the objects in the region – cannot lack significance for the exploration of empirical facts.

Therefore, when we consider the operations of transition to species or genera (specialisation and generalisation), we are in the field of properly ordered essences – from the highest to the lowest genus. Again, let us give the floor to Husserl himself [8, p. 25].

> We now need a new group of categorial distinctions pertaining to the whole sphere of essences. Each essence, whether materially filled or empty (thus, purely logical), has its place in a hierarchy of essences, in a hierarchy of *generality* and *specificity*. This series necessarily has two limits which never coincide. Descending, we arrive at the *infimae species* or, as we also say, the *eidetic singularities*; ascending through the specific and generic essences, we arrive at a *highest genus*. Eidetic singularities are essences which necessarily have over them "more universal" essences as their genera, but do not have under them any particularizations in relation to which they would themselves be species (either proximate species or mediate, higher, genera). In like fashion, that genus is the highest which has no genus over it.

Let us now present a concrete example of hierarchically ordered essences. Referring to the studies of philosophers, we can indicate the well-known Porphyry tree (*arbor porphyriana*) [10, 23].

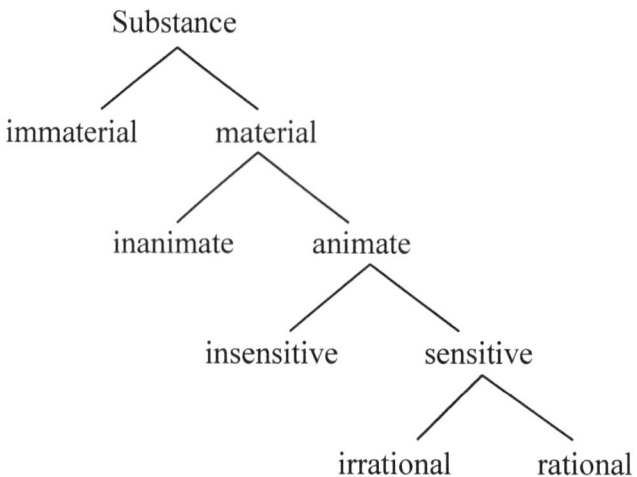

Let's also establish that:

(a) the substance (as a universal) is characterised by its content (ideal quality, in Ingarden terminology): "being a substance",
(b) material substance by: being a substance and being material (I omit quotation marks),
(c) immaterial substance by: being a substance, being immaterial; example: angel,
(d) sensitive substance by: being a substance, being material, being sensual,
(e) rational substance by: being a substance, being material, being sensual and being rational.

I skip the description of the other objects, because it is easy to guess. Furthermore, let us notice that an irrational substance could be replaced by many other "objects" such as "equines", "elephants", etc. For example, in zoology, a horse is characterized as: multicellular, vertebrate, mammal, and odd-toed (in short); let us treat it as having the following content: being a substance, being material, being sensual, being odd-toed.

Next, the particular names in the Porphyry tree should be treated as names for so-called universal objects. If there are some dashes down from a certain inscription, this inscription is the name of the genus, and if there is nothing underneath, this is the name for the species. Thus, when Husserl speaks of the lowest varieties of universal objects, he indicates the species (not the genera). So if a human being (a rational substance) is a species, or the lowest kind, then there is no such thing as a species or essence: male, female, hairdresser or philosopher. Species (but also genera) are sometimes called essences by Husserl (the Greek term *eidos* is sometimes translated as idea, sometimes as essence). Specialisation is the transition from a genus (e.g. animate substance) to a lower genus or species (e.g. to a sensitive substance or immediately to a rational substance). Generalisation goes in the opposite direction (e.g. from a human being to an animate or material substance).

Things are obvious when we have a tree. But how do we get it? Let us notice that also the above tree can "miss" essences, although the philosophical tradition convinces us that e.g. "animality and rationality" is the essence of man. Zoologists and philosophers build different "systematics" of animals, plants and man (one of the animals). The aim of Husserl is therefore to bring out what is the essence of what is alive, what is the essence of man, and so on. In his Ideas (that is, in Volume II) he gives, among other things, an answer that can be given briefly as follows: the essence of an organic substance is: being a substance and being alive (of course, we could discuss both at length). What is more, I would also like to stress that the transition from a certain kind, to a kind that is directly inferior (e.g. from what is animate to what is sensitive) does not have to be made by indicating a single content. Content: "being sensitive" or "being reasonable" are usually very complex contents.

Ingarden understood these species and genera (he called them, in general, ideas) as follows (I will give it by example and in a formalized way).[1] In the material substance as such we have certain contents, let there be five of them, from u_1 to u_5, which together define what is substantiality (being a substance). Furthermore, we have, let us say, four contents, let us mark them with the letter w in the appropriate indexes, which characterize what we briefly express as "materiality". This is not all, because in such an idea there are still – according to Ingarden – some variables, i.e. other contents, but not yet defined, and which concern organicity (the letters x), sensuality (y) or rationality (z). If we define the letters x in the appropriate indexes negatively, we obtain an inorganic material substance, an example of which is stone, while if we define the letters x and y positively, we obtain the idea of material substance, organic and sensual. However, a problem arises: can we talk about a material, inorganic and sensitive substance? Is there such an idea, such an essence? Well, here is the biggest problem that the philosopher is trying to solve. Husserl's answer, and Hartmann in particular later, goes in this direction to discover that "there is no sensuality without organicity". It is true that Thomas Aquinas taught about angels, which were immaterial and rational substances, but in our real world, rational beings (man) are only those which by necessity must also be: material, organic and sensitive (let us note that Kant has already taught that all cognition begins with intuition, with sensuality), so without senses there would be no reason, and without organic there would be no senses.

After these explanations, it is clear that specialization is the transition from a higher order essence to a lower order essence. But: not blindly! Not everything is an essence, not every filling with the contents of a higher essence hits a lower kind of essence. For example, there is no such thing as a material, inanimate, insensitive and rational substance[2]. Generalization in turn is the reverse process. But also here we can see that if we take the essence of the human being, we cannot make any content variable (inverse to filling it with content), e.g. (the answer is partly in the language of science) we cannot move from the idea of the human being to the idea of something that does not have a nervous system or is not a vertebrate, although it remains (sic!) reasonable.

Remark. The Porphyry tree is a good example of classification or so-called logical partition. The classification assures us that by distinguishing certain subgenera, we distinguish those subgenera whose subordinate individuals are all individuals of a given type, and those subgenera are such that the subordinate individuals do not simultaneously fall under other subgenera. However, the following problem arises: when we distinguish in a kind of polygon such as the regular and non regular polygon or the concave and convex polygon, which of these partitions is appropriate? Which of these partitions "hits" the essence? Of course, mathematicians are not interested in such problems today. It is a philosophical problem. A mathematician is interested in concepts (or mathematical structures and objects), a philosopher is interested in essences.

Let us now move on to the next pair of operations: formalization and de-formalization. These are operations different from the specialisation and generalisation operations just discussed. In the Paragraph 13 *Generalization and Formalization* Husserl explicitly states [8, p. 26]:

> One must sharply distinguish the relationships belonging to generalization an specialization from the essentially heterogeneous relationships belonging, on one hand, to the *universalization of something materially filled in the sense of pure logic* and, on the other hand, to the converse: the *materialization* of something logically formal. In other words: generalization is something totally different from that *formalization* which plays such a large role in, e.g., mathematical analysis; and specialization is something totally different from *de-formalization*, from "filling out" an empty logico-mathematical form or a formal truth.

Husserl explains these difficult operations (formalization and de-formalization) by analysing examples from the field of mathematics (geometry) and the sphere of sensual quality [8, p. 26].

Accordingly, the subordinating of an *essence* to the formal universality of a *pure-logical* essence must not be mistaken for the subordinating of an essence to its higher essential *genera*. Thus, e.g., the essence, triangle, is subordinate to the summum genus, Spatial Shape; and the essence, red, to the summum genus, Sensuous Quality. On the other hand, red, triangle, and similarly all other essences, whether homogeneous or heterogeneous, are subordinate to the categorical heading "essence" which, with respect to all of them, by no means has the characteristic of an essential genus; it rather does *not* have that characteristic relative to *any* of them. To regard "essence" as the genus of materially filled essences would be just as wrong as to misinterpret any object whatever (the empty Something) as the genus with respect to objects of all sorts and, therefore, naturally as simply the one and only summum genus, the genus of all genera. On the contrary, all the categories of formal ontology must be designated as eidetic singularities that have their summum genus in the essence, "any category whatever of formal ontology."

Apart from explaining what formalizing and de-formalizing is, Husserl points out the differences of the above operations in relation to the operations of generalization and specialization. Nevertheless, let us give some more examples from philosophical fields.

1) In the Porphyry tree, we have indicated specific materially defined essences. Ingarden, as I wrote above, understands them properly. Note that each essence has a certain amount of content that has appeared at a higher level and a new set of content that appears as a filling of the higher level. The latter set is that which in scholastics corresponds to the species difference, the former to the directly superior genus. Well, we can say that when we consider an essence (universal object) as an empty thing, we are not interested in material terms, but only in the pure form of the essence, in which we discover the "generic part" and the "species difference part". This is formalization!

2) Let us consider the following reasoning (argumentation):
(A) If the cube of sugar is placed in boiling water, then the cube will dissolve

And

The cube was placed in boiling water,

Thus

The cube will dissolve,

This is an example of some detailed (material) reasoning. But when the logician comes to the conclusion that the general scheme of this inference is a formula

(*) $((\alpha \to \beta) \wedge \alpha) \to \beta,$

we have an example of formalization. Of course, the formula (*) is not any genus (kind) in relation to reasoning (A). Husserl explains it as follows [8, pp. 26-27]:

> It is clear, similarly, that Any determinate inference, e.g., one ancillary to physics, is a singularization of a determinate purely logical form of inference, that any determinate proposition in physics is a singularization of a propositional form, and the like. The pure forms, however, are not genera relatively to the materially filled propositions or inferences, but are themselves only infimae species, namely of the purely logical genera, proposition, inference, which, like all similar genera, have as their absolutely highest genus "any signification whatever".

3) In the monograph [11] I recalled Wolff's views on being. For Wolff, being is what is non-contradictory, what is possible. Every being is determined by the essential, attributive and contingent features (properties). It is usually stated in the philosophical literature that organicity, animality or rationality are examples of essential qualities. Then the attributes will be the ability to use language or create knowledge, while the contingent features will include being a philosopher or having two children. However, when we point to such features of particular entities or classes of entities, then we are in the

area of material, regional ontology. An important result of Wolff's ontology, however, is that he formalized the concept of being. How did he do this? He did it by indicating three classes of properties and establishing mutual relations between them. For example, essential properties are independent of each other, attributive properties are generated by essential properties, while contingent features are those that are inconsistent with essential properties. These relationships and their properties apply to each material domain and are independent of each domain. Therefore Wolff gave a formal approach to being, and the transition from these and these material domains (e.g., from animal existence) to the formal approach of being is a formalization (compare details of this analysis in [11, pp. 40-43] and [28]). In turn, the transition from a formal approach of being to an animal or human being, which is not easy and is done as a result of proper filling with content, is what Husserl calls a de-formalization operation.

3. Topological Hermeneutics

In this chapter I would like to draw attention to the topological ontology that has been developing in recent years and its method, which I call topological hermeneutics. Topological ontology (in short topoontology) as a fragment of topological philosophy is an analysis of ontological concepts, assumptions, theorems and problems using concepts, statements and tools of general topology. This kind of analysis has been undertaken in the works of Mormann [17], Schulte and Cory [19], Skowron [20], [22], and Kaczmarek [13], [14], [15]. What is topoontology and what is topological hermeneutics? I will explain this, I hope, more fully when I present particular ontological solutions using general topology tools.

I compare the study of ontology problems using topological tools with the studies of Wolniewicz, who presented a precise interpretation of Wittgenstein's ontology by applying the lattice theory (comp. Wolniewicz [30] and [31]). What is more, Wolniewicz proposed the so-called logical hermeneutics, which allows for the interpretation and comparison of certain theses of Wittgenstein's ontology and Hume's epistemology in the lattice theory[3]. My proposal is to use a general topology to interpret Wittgenstein's ontology, Hume's epistemology and Leibniz's ontology (monadology). It turns out that the Wolniewicz's lattices can be understood as lattices composed of certain topological spaces and thus we obtain a generalisation of Wolniewicz's theory. Topological hermeneutics therefore concentrates on the fact that it incorporates various notions and theorems of ontology in the language of general topology and not (only) in the language of the lattice theory or logic. In my opinion, as I will try to demonstrate, such an approach results in new and interesting formal theorems that have ontological significance. So let's move on to the concrete ones. I will focus mainly on the topological interpretation of small fragments of Wittgenstein's logical atomism ontology (and Russell's, because they worked on these issues together).

We will conduct our considerations on the example of two lattices examined by Wolniewicz in [30, p. 81]: the first lattice is an atomic lattice with W-independent elements, the second is a non-atomic lattice with W-independent elements.

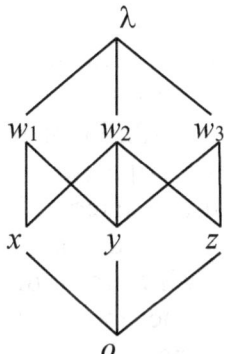

Figure 1. Atomistic lattice.

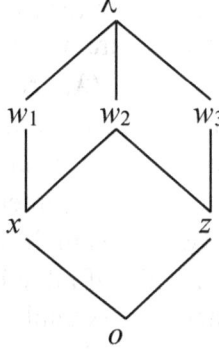

Figure 2. Non-atomistic lattice.

The elements of these lattices are interpreted as situations: o is an empty situation, λ is an impossible situation and the others are proper situations. Situations x, y and z are atomic and correspond to the Wittgenstein's states of affairs. In turn, w_1, w_2 and w_3 are called possible worlds, and we can interpret them as conjunction (splice, concatenation) of atomic situations.

Before we move on to further considerations, let us explain three concepts: atomistic lattice, non-atomistic lattice and W-independence of situations. The concept of the atomic lattice – different from the concept of the atomistic lattice – and concept of topological space – will also help. Definitions of these concepts can be given in purely formal language (in the language of lattice theory). However, we will abandon this way of defining and present these definitions in natural language (using maximum precision).

There is a certain order < in each lattice K. For example, in the Lattice from diagram 2: $x < w_1$ and $x < w_2$. The smallest element o is called a zero of the lattice, and the largest element λ is called a unity of the lattice. For any $a \in K$ and $a \neq o$, the set $[o, a] = \{ x \in K : o < x < a \}$ is called a segment. The element a (different from zero) of the lattice is called an atom if the segment $[o, a]$ is two-element one.

1) a lattice K is atomic iff in any interval $[o, a]$ there is an atom; as you can see, both lattices above are atomic;

2) a lattice K is atomistic iff each element of the lattice is the supremum of some set of atoms; in the above examples, the first lattice is atomistic and the second is not; for example, in figure 2, element w_1 is not the supremum of any set of atoms;

3) two elements x, y of the lattice K are called W-independent (Wittgenstein's concept of independency) iff infimum of x and y is o whereas supremum x and y is different from λ; for example, element x i y of Figure 1 are independent, but x and w_3 are dependent;

4) if X is Any set, then the pair (X, τ_X) will be called topological space, where τ_X is Any family of subset of X iff the family fulfils the following conditions: a) the empty set $\emptyset$ and X belong to τ_X, b) any union of subsets of X belongs to τ_X and c) intersection of finite number of subsets of X belongs to τ_X; an example of a topological space is a pair (X, τ_X), where τ_X is a family of all subsets of a set X; this space is called discrete space; another example is the so-called Euclidean space on a set of real numbers R, where τ_R is composed of

sets which are the union of any number of intervals (a, b), for $a, b \in$ **R**.

It turns out that the above presented lattices can be transformed into lattices composed of topological spaces. I then propose the following procedures for conversion. In Figure 1, we convert:

o into $\emptyset$,
x, y and z to (respectively) $\{x\}$, $\{y\}$ and $\{z\}$,
w_1, w_2, w_3 we convert into $\{x, y\}$, $\{x, z\}$ and $\{y, z\}$.

Then it is easy to see that e.g. the family of sets included in w_2 i.e. the family $\{\emptyset, \{x\}, \{z\}, \{x, z\}\}$ together with the set $\{x, z\}$ is a topological (discrete) space, and the appropriate lattice can be visualized as follows:

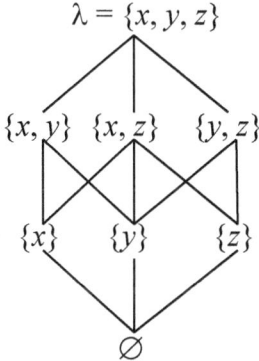

Figure 3. Atomistic lattice with three topological spaces.

We do the same with the lattice presented in Figure 2. Here, however, both w_1 and w_3 are not suprema of the selected group of atoms and therefore we have to propose that w_1 we convert to $\{x, y_1\}$ and w_3 to $\{y_2, z\}$. Then we again see that e.g. the family of sets included in w_1 i.e. the family $\{\emptyset, \{x\}, \{x, y_1\}\}$ together with the set $\{x, y_1\}$ constitutes a topological space. It is easy to see that this space is not discreet.

The above procedure allows us to obtain an interesting topoontological statement. Namely,

Fact. Any atomic lattice is atomistic when it is composed of discrete topologies.

In this way we received the necessary condition for each element in the atomic lattice to be the supremum of a certain set of atoms (in the language of general topology: that each set is the union of a certain set of singletons). Atomicity and atomisticity are, according to Wolniewicz, the key assumptions of Wittgenstein atomism. Following Wolniewicz, we can say that every possible world, including our real world, according to *Tractatus*, can be interpreted as a multiplicity or total of all atomic states of affairs that are *W*-dependent.

In the paper [14] I also considered non-atomic lattices, i.e. ones which do not meet the condition that in any segment an atom exists. Is it worth to consider such lattices? Well, Wittgenstein assumed that the analysis of a sentence cannot be carried out indefinitely, so there must be so-called elementary sentences and consequently their correlations on the side of reality, i.e. atomic states of affairs. However, when asked about an example of a simple sentence that refers to an atomic state of affairs, he replied that he did not know. Nota bene in *Tractatus* we will not find such an example either. The problem is that a simple sentence of the type

'The weather is nice'

can be seen as a conjunction of sentences

'It's sunny and warm.'
But then a simple sentence

'It's warm'

we can interpret as a conjunction of sentences, say,

'It's such a such temperature and it doesn't blow'

and, theoretically, we can further analyze other simple sentences (e.g. 'It is sunny' we can describe by the state of cloudiness and type of clouds). This makes us think that it is worthwhile to study such lattices, in which a given situation (in topology a certain set) can be analyzed by smaller sets, e.g. $A = B \cup C$, next $C = D \cup E \cup F$, and thus $A = B \cup D \cup E \cup F$, and so on. The use of topological spaces allows for the interpretation (modelling) of both atomic and non-atomic theses.

In this paper I also took up another problem that was suggested by P. Weingartner: what is the negation of the atomic state of affairs and is it also an atomic state of affairs? It turns out that the answer is the following:

a) in Wolniewicz's lattices, the negation of an atomic state of affairs may be another atomic state of affairs or, also, a complex situation (consisting of several states of affairs); let us refer to Figure 1; elements x and z are atoms, w_3 is not an atom and is the supremum of y and z; it turns out, however, that the infimum of x and z is the zero of the lattice, while the supremum of x and z is the unit of the lattice, which means that z is the negation of the atom x; the same is true for the x and w_3; their infimum is zero and the supremum is the unit of the lattice; conclusion: w_3 is also the negation of x; the negation of x is therefore both the atomic and the complex element,

b) another result is obtained in the case of non-atomic lattices; in [14] I showed that there are lattices consisting of topological spaces, in which for any situation (a set) there is no negation of it (complement of such a set is not a part of the lattice)[4]. Ontologically we can interpret this result as follows: when we consider possible worlds, including our real world, all situations or states of affairs are positive. No situation is a negation of any other. This answer is consistent with the theses of those ontologists who doubt the existence of negative states of affairs or negative situations.

4. Summary and Final Remarks

In this piece I tried to show that the methods of analytical philosophy indicated by Woleński can be supplemented. After all, a few decades have passed. So I added the methods or operations proposed by Husserl and presented briefly the method (or tool) called here topological hermeneutics. I hope that Professor Woleński will agree with this proposal.

Let us try to sum up: what is topological hermeneutics as a method or a certain tool within an analytical method? Ontological hermeneutics is doing so:

1) considers the problems of classical ontology (e.g. the main theorems of logical atomism (among others, atomicity), what is a monad (Leibniz's ontological atomism), what are perceptions and how they relate to the situation (Hume, Wolniewicz's logical hermeneutics)),

2) formalises the theses (but also concepts) studied in the language of general topology, because, in the case of the interpretation of logical atomism in Wolniewicz's view, it turns out that this interpretation can be generalized to study both the approach characteristic for Wittgenstein's and Russell's atomism and the approach opposite to atomism,

3) derives formal theses concerning atomism and non-atomism in the language of general topology,

4) leads to new conclusions which cannot be proved on the basis of the theory of Wolniewicz's lattices (cf. Fact given above); these conclusions shed new light on the situation ontology and logical atomism,

5) derives formal theorems, which can be interpreted ontologically, but also, and we hope so, can influence the search for mathematicians themselves.

There is one more problem that I have set myself as a task for the future. It is about the validation (Ajdukiewicz's term) or justification (Bocheński's term) of the operations, tools, methods discussed. In the case of justification I think that points a) – d) indicated in the final part of Paragraph 1.2 of this paper can be accepted as justification for the topological hermeneutics method. Perhaps we should look for more justifications. However, in the case of Ajdukiewicz the matter is slightly different. Ajdukiewicz tries to find a certain logical theory (a certain set of logical sentences) that would be the basis for philosophical claims. This basis would guarantee the validation of philosophical theorems (which are usually given in natural language). Ajdukiewicz did not see a solution when he was writing about it, and I do not see a solution today either. This should be put as a problem. I think it is a key problem. We may ask: for which philosophical field is it a key problem? The short answer is: for everyone who considers the results of formal sciences (these, according to Aristotle, were a tool of philosophy). So let it be a problem which will be dealt with by ontologists and logicians.

Acknowledgments

This paper is supported by the National Science Centre, Poland, No 2017/27/B/HS1/02830.

References

1. Ajdukiewicz, K. On The Applicability of Pure Logic to Philosophical Problems, In K. Ajdukiewicz *The scientific world-perspective and other essays, 1931-1963*, Synthese Library, vol. 108, Dordrecht, Holland: D. Reidel Publishing Company, 1978 (1934), pp. 90-94.
2. Ajdukiewicz, K. A Semantical Version of the Problem of Transcendental Idealism, In K. Ajdukiewicz *The scientific world-perspective and other essays, 1931-1963*, Synthese Library, vol. 108, Dordrecht, Holland: D. Reidel Publishing Company, 1978 (1937), pp. 140-154.
3. Ajdukiewicz, K. *The scientific world-perspective and other essays, 1931-1963*, Synthese Library, vol. 108, Dordrecht, Holland: D. Reidel Publishing Company, 1978.
4. Bocheński, I. M. *The Methods of Contemporary Thought*, New York: Harper Torchbooks, 1968 (first ed. Die zeitgenössischen Denkmethoden, Dalp TB, Bd. 304, Bern: Francke, 1954).
5. Czeżowski, T. *Logika. Podręcznik dla studiujących nauki filozoficzne (Logic. Handbook for Students of Philosophy)*, Warszawa: Państwowe Zakłady Wydawnictw Szkolnych, 1949.
6. Glock, H-J. *A Wittgenstein Dictionary*, Oxford: Blackwell Publishers, 1996.
7. Hempel, C. G., H. Putnam, and W. K. Essler (eds.). *Methodology, Epistemology, and Philosophy of Science: Essays in Honour of Wolfgang Stegmüller on the Occasion of his 60th Birth Day, June 3rd, 1983*, Reprinted from the Journal *Erkenntnis* 19 (1, 2 and 3), Springer Verlag, 1983.
8. Husserl, E. *Ideas Pertaining to a Pure Phenomenology and to a Phenomenological Philosophy*, trans. by F. Kersten, The Hague/Boston/Lancaster: Martinius Nijhoff Publishers, Kluwer Academic Publishers Group (first edition: 1913), 1983.
9. Ingarden, R. *Controversy over the Existence of the World*, Vol. I, Translated and annotated by A. Szylewicz, Frankfurt am Main: Peter Lang GmbH, Internationaler Verlag der Wissenschaften, pp. 320 (first edition: Kraków: PAU, 1947/48), 2013.
10. Kaczmarek, J. On the Porphyrian Tree Structure and an Operation of Determination, *Bulletin of the Section of Logic* 31/1, 2002, pp. 37-46.

11. Kaczmarek, J. *Indywidua. Idee. Pojęcia. Badania z zakresu ontologii sformalizowanej* [*Individuals. Ideas. Concepts. Investigating into Formalised Ontology*], Łódź: Wyd. Uniwersytetu Łódzkiego, 2008.

12. Kaczmarek, J. What is a Formalized Ontology Today? An Example of IIC, *Bulletin of the Section of Logic* 37 (3-4), 2008, pp. 233-244.

13. Kaczmarek, J. Atom ontologiczny: atom substancji [Ontological atom – atom of substance], *Przegląd Filozoficzny. Nowa Seria* R. 25, nr 4 (100), 2016, pp. 131-145.

14. Kaczmarek, J. Ontology in Tractatus Logico-Philosophicus: A Topological Approach, In G. Mras, P. Weigertner, and B. Ritter (eds.), *Philosophy of Logic and Mathematics*, Berlin/Boston: De Gruyter, pp. 245-262, 2019.

15. Kaczmarek, J. On the Topological Modelling of Ontological Objects: Substance in the Monadology, In B. Skowron (ed.), *Polish Contemporary Ontology*, Berlin/Boston: De Gruyter, pp. 149-159, 2019.

16. Kuratowski, K. *Wstęp do teorii mnogości i topologii, (wraz z dodatkiem R.Engelkinga: Elementy topologii algebraicznej)*, [*Introduction to Set Theory and Topology* (*with a supplement by R. Engelking: Elements of algebraic topology*)], Warsaw: PWN, (see also: Kuratowski, *Topology*, vol. I, 1966, vol. II, 1968), 1977.

17. Mormann, T. Topology as an Issue for History of Philosophy of Science, In *New Challenges to Philosophy of Science* Volume 4 of the series *The Philosophy of Science in a European Perspective*, pp. 423-434, 2013.

18. Porphyry. *Introduction (or Isagoge) to the logical Categories of Aristotle*, vol. 2, pp. 609-633, (this text was transcribed by Roger Pearse, Ipswich, UK, 2007), see: http://www.tertullian.org/fathers/porphyry isagogue 02 translation.htm#c, 1853.

19. Schulte, O., and J. Cory J. Topology as Epistemology, In B. Smith and W. Żełaniec, Topology for Philosophers, *The Monist* 79 (1), 1996, pp. 141-147.

20. Skowron, B. The Forms of Extension, In M. Szatkowski M. and M. Rosiak (eds), *Substantiality and Causality*, Philosophische Analyse/Philosophical Analysis, Berlin/Boston: De Gruyter, 2014, pp. 175-187.

21. Skowron, B. Using Mathematical Modeling as an Example of Qualitative Reasoning in Metaphysics. A Note on a Defense of the Theory of Ideas, *Annals of Computer Science and Information Systems* 7, pp. 65-68, 2015.
22. Skowron, B. Mereotopology, In J. Seibt, S. Gerogiorgakis, G. Imaguire, and H. Burkhardt (eds.), *Handbook of Mereology*, Philosophia, München: De Gruyter, pp. 354-361, 2017.
23. Strange, S. K. *'Introduction' to Porphyry: On Aristotle's Categories*, London: Bristol Classical Press, 1992
24. Wittgenstein, L. *Tractatus logico – philosophicus*, (first ed. [1921], *Logisch – philosophische Abhandlung*, In *Annalen der Naturphilosophie*, Warszawa: BKF, PWN, 1997.
25. Woleński, J. Kierunki i metody filozofii analitycznej [Directions and Methods of
Analytical Philosophy], In J. Perzanowski (ed.), *Jak filozofować? [How to Philosophise?]*,
Warszawa: PWN, pp. 30-77, 1989
26. Woleński, J. *W stronę logiki [Towards Logic]*, Kraków: Wyd. AUREUS, 1996.
27. Woleński, J. *Epistemologia [Epistemology]*, Warszawa: Wyd. Naukowe PWN, 2005.
28. Wolff, Ch. *Philosophia prima sive Ontologia metodo scientifica pertractata qua omnis cognitionis humanae principia continentur*, Veronae (comp. Pars I, Caput III: *De notione entis*), 1789.
29. Wolniewicz, B. Hermeneutyka logiczna [Logical Hermeneutics], *Studia Filozoficzne* 7, pp. 27-40, 1983.
30. Wolniewicz, B. *Ontologia sytuacji [Situation Ontology]*, Warszawa: PWN, 1985.
31. Wolniewicz, B. *Logic and Metaphysics. Studies in Wittgenstein's Ontology of Facts*, Warszawa:
Polskie Towarzystwo Semiotyczne (Ed. by Polish Semiotic Association), 1999.

Notes

1. The Reader can find Ingarden's investigations on ideas in Ingarden [9], Chapter II, § 9 and also in other chapters.
2. This is the case, for example, according to Hartmann and – probably – is confirmed by the science of facts (to follow Husserl's terminology). However, as philosophers, we cannot insist on such a

position. Personally, I think that when, for example, angels are said to be immaterial and rational, the term "rational" means something different from the human being defined as *animal rationale*.

3. Comp. [29]. Wolniewicz writes in the abstract of his paper: "Rules and evaluation criteria for the interpretation of philosophical systems are called hermeneutics. The logical interpretation of a system is aimed at revealing its logical structure. Its hermeneutical value depends on several parameters: range, coherence, naturalness, additional assumptions, and concordance with other systems. For illustration purposes, significant fragments of two known metaphysical systems were interpreted in this way: Hume and Wittgenstein."

4. Formal details and a discussion of these issues can be found in [14, pp. 412-414], while the definition of a lattice composed of topological spaces can be found on p. 405.

Anti-foundationalist Philosophy of Mathematics and Mathematical Proofs

Stanisław Krajewski

University of Warsaw
Krakowskie Przedmieście 3 Street
00-927 Warszawa, Poland
e-mail: stankrajewski@uw.edu.pl

1. Historical Background: from Euclid to Hilbert

For centuries mathematical proofs have been seen as special, different from any other kind of argument. Mathematicians and all educated Westerners could point to their exceptional traits: proofs in mathematics seem more precise, more elaborate, more compelling, more certain, more logical than any other proof-like discourse – so much more that they can be seen as absolute. A crucial evidence has been provided by the Euclidean axiomatic system of geometry. This book was taught to all who were able to follow mathematics and served as a paradigm of mathematical argument. Euclid's system was seen as complete, all geometrical theorems were supposedly reducible to the initial general "common notions" and specific postulates. As late as the 19th century, it turned out that some implicit assumptions were used and that a more complete treatment was needed in order to achieve the goal of having the system of geometry that is purely logical and does not depend on intuitive visualization. This was possible due to the work of Moritz Pasch and David Hilbert. In addition, the development of non-Euclidean geometries showed the limitations of the intuitive methods and the need for rigor. All of these developments did not diminish the influence of the Euclidean ideal of axiomatic mathematics. Rather, they seemed to confirm the view that mathematics consists, at least ideally, of axiomatic theories that can be presented in a very rigorous way, making explicit all assumptions.

One element of the contemporary version of the axiomatic method has been different from the approach of Euclid: rather than defining directly the objects of the theory (for example, points and lines) the objects were indirectly defined by the axioms that expressed the main properties of the objects and, even more important, basic relations between the objects. Nothing more was assumed than what was stated by the axioms. Hence Hilbert's famous remark that the objects of his system of geometry can be anything, for instance "tables, chairs, and beer mugs," as long as they satisfy all the axioms. This approach made possible a new variant of the axiomatic method; it slowly emerged in the 19th century. Namely, arbitrary axioms can be proposed and their realizations studied. Hence the notion of a group and other structures studied in abstract algebra. How they can be applied to the world is another matter. Pure mathematicians may disregard it. In practice, however, axioms were never completely arbitrary; rather, they conveniently codified regularities observed in the world of mathematical objects. Yet the idea that axiomatic theories can have multiple realizations became a new norm. In the 20th century the theory of models emerged, or a study of possible theories and their various interpretations.

In order to have a strict mathematical theory of models it was necessary to have a full description of the logical machinery utilized to prove theorems form axioms. This was possible due to the work of Frege and later proponents of logicism. Hilbert was happy that as if in result of "a preestablished harmony" logic itself was axiomatized: the so-called first order logic was identified as basic.

In addition, due mainly to Georg Cantor, actually infinite sets were introduced as an object of study in mathematics. The general concept of a set was also necessary in order to develop systems of higher order logics that reflected methods naturally used by mathematicians. To make clear what properties of sets may be used so that we can avoid antinomies that were plaguing the early research dealing with infinite sets, Zermelo axiomatized set theory. Since then, in the early 20th century, it was developed by Fraenkel and others so that the ZF (or ZFC, that is, ZF with the axiom of choice added) system emerged that has been seen as an adequate basis for abstract mathematics. Interestingly, the axiomatization of set theory was made in the spirit of Euclid: the principal properties of the intuitive concept of a set were listed so that all other properties of "pure sets" could be logically derived.

As a result of all those well-known developments, some hundred years ago it became widely agreed that the axiomatic method could be seen as normative. Its strengthening, namely the notion of a formalized theory, became the ideal of mathematical theory, especially for those who assumed that the right approach to mathematics must be grounded in logic. A formalized theory is axiomatic, the axioms are expressed in a perfectly defined language, its underlying logic is axiomatized, and the meanings are assumed to be grasped by all these axioms together with formal rules of derivation of formulas from other formulas. This picture of the axiomatic approach and its refinement, the notion of formal theories, has been highly successful and extremely influential among philosophers. For some analytic philosophers this picture became a model of scientific and even philosophical analysis.

The notion of axiomatic mathematics involved an understanding of mathematical proof. Its essence was seen in Hilbert's concept of formal proof: it is a sequence of formulas of the underlying formal language, each of the terms of the sequence being either an axiom or the result of an application of one of the explicitly listed formal rules of inference to previous terms of the sequence. There are variants of these notions, for example the sequent calculus, and extensions, for example rules with infinitely many premises, but the general idea remains: proofs are essentially derivations, very much like calculations. While everybody knows that real proofs are very different from this ideal the supposition was that they are humanly available indications of ideal proofs. The underlying assumption, then, called sometimes Hilbert's Thesis or the Frege-Hilbert Thesis, is as follows:

> Every real mathematical proof can be converted into a formal proof in the appropriate axiomatic theory.

This attractive hypothesis has been, however, rejected by more and more philosophers of mathematics since at least the 1960s.

2. Movement Against the Euclidean Notion of Proof

Probably most mathematicians do not really care whether real proofs can be converted to formal proofs or not. They may believe those colleagues who say that this is the case, but they know well that this has nothing to do with their practice of proving mathematical results. Many would probably express doubts as to whether the formal proof is really always possible, even in principle. It is hard for me to say how many would, since I have not heard about representative studies on the issue conducted among professional mathematicians.

Whatever the opinions regarding Hilbert's Thesis among those who produce proofs, an increasing number of philosophers of mathematics and mathematicians reflecting upon their profession have begun to analyze mathematical proofs as they really are. This is a part of a more general turn in the philosophy of mathematics. The change began with the analysis of proofs of Euler's formula for polyhedral, $V-E+F=2$, made brilliantly by Imre Lakatos in the 1960s. Among others who contributed to the new trend let me mention Philip Kitcher, Reuben Hersh, Paolo Mancosu, Yehuda Rav, Carlo Cellucci, Brendan Larvor, David Corfield, and Brian Rotman. Their positions on many issues in the philosophy of mathematics differ, but all tend to deny the possibility of, and the need for, foundations of mathematics, that is, the idea of reducing the whole of mathematics to one theory, treated as its foundation. This new attitude is sometimes called, after Aspray and Kitcher [1, p. 17], "the maverick" tradition. It is opposed to the traditional philosophical schools of the foundations of mathematics: logicism, formalism, constructivism (including intuitionism). Some representatives of the new approach are playing down the role of logic. Many want to understand mathematics as a part of human culture. Most of them doubt, to varying degrees, the adequacy of realism in the philosophy of mathematics. All want to begin with genuine mathematical practice.

It will be useful to mention briefly some of the main points made in their works, especially those that are relevant to the analysis of proofs. I will summarize some views of a few of the above-mentioned authors, those who according to me have been most innovative. Actually, there is something paradoxical in looking for novelty in this new approach to mathematics, as the point of the new trend was to observe closely what real mathematicians actually do rather than to invent something new about them. A tension is, however, inevitable between experiencing, in this case experiencing mathematics, and describing the experience. We always need to indicate what strikes us as most important and name it, and this often requires invention: we try to detect relations, which may be hidden; we attempt to form a picture of the mechanism underlying the experience; and it may happen that we become aware of the realities that are so obviously present as to be missed in earlier descriptions. (See below, in this section, examples of each of these three categories: (i) hidden relations, (ii) underlying mechanisms, (iii) obvious features that are easily ignored.) More generally, we never provide a completely neutral account of an experience or a historical process, even if we do our best to remain neutral. Rather, we present a reconstruction taking advantage of our understanding of the situation. In the case of mathematics this can be far from obvious.

Thus, Lakatos in his celebrated book [25], based on papers written in the 1960s, presented the theory of the dialectical process of the development of mathematics from proof to refutation to improved proof to another refutation, etc. This means that proofs can be mistaken or at least imperfect even if they are recognized as flawless. The refutation comes from the (intuitive) mathematical background that provides potential falsifiers. By the way, Lakatos provided an insightful rational reconstruction of the historical process of proving, so this is an example of (ii), the underlying mechanism of the mathematical experience, namely the process of proofs and refutations. Also, he indicated the relation of proofs to the environment in which they live, and which can provide counterexamples. Lakatos introduced the term "quasi-empiricism" (see his [26]) together with the claim that the methods used to establish results in mathematics are not as (qualitatively) different form natural sciences as had been assumed in the received tradition in the philosophy of mathematics. (The term "quasi-empirical" was also used by Putnam [30].)

Reuben Hersh, generally known for a beautiful popularization of mathematics – the real one, not the logicians' picture of it – in the book [7], co-authored with Philip Davis, is another forefather of the maverick tradition. In [16] he introduced the distinction between the front and the back of mathematics. This distinction, borrowed from sociological and cultural studies is, by the way, a good example of (iii), an obvious feature that was ignored by philosophers of mathematics. Namely, it is clear to every mathematician that official mathematics, presented in publications and formal lectures, is radically different from the tentative efforts, guesses, trials, hypotheses and mistakes present in the mathematical kitchen. Hersh also advocated, on many occasions, the idea that mathematical entities are cultural creations having an intersubjective reality. This cultural approach was initiated by Raymond Wilder [43] (see also [44]), but Hersh was emphasizing much more strongly the inflexibility and objectivity of mathematical creations, another point obvious to any working mathematician.

Let me mention that to represent both aspects, createdness and objectivity of mathematical entities, and keep them as equally important I have introduced the concept of "suprasubjective existence" in [24]. Suprasubjective is defined as intersubjective and, at the same time, "objective without objects."

Rav [32] argues that many mathematical theories have not been axiomatized and it seems that they will never be: any attempt to do this would require far reaching changes in the theory. Even group theory, defined by axioms of the group, uses higher order methods that have little to do with axiomatic theories. And actually there has never "been a unique conception what axioms are" [33, p. 125]. Independently of this, Rav [31] proposed an interesting solution to the age old problem of whether what we do in mathematics can be characterized as invention or discovery. According to his proposal, concepts are invented and theorems are discovered. In relation to our main topic, he emphasized the crucial role of proofs in mathematics. They are the heart of the matter. Theorems are only convenient expressions of what has been or can be proved. Proofs are like bus routes and theorems like bus stops that are established in a rather arbitrary way.

Cellucci, in several publications, for example in [4] and in [6], has been advocating the concept of analytic proof that he traces back to Plato, while the concept of axiomatic proof, used by Euclid, was recommended by Aristotle. Cellucci reminds us that a mathematical work begins not with axioms but rather with a problem. To produce an analytic proof one has to find a suitable hypothesis that makes it possible to solve the problem. This hypothesis must be plausible and sufficient for a derivation of the theorem. The derivation may be deductive, but this is not necessary. Thus the crux of the proof is to find the suitable hypothesis. It may be a construction, a concept, a theorem, a picture, a theory, or a conjecture. The search for a right hypothesis is certainly pervasive in research and this, by the way, provides an example of (i), a hidden relationship between elements of mathematical experience. Cellucci claims that everything in mathematics is hypothetical: concepts, objects, theorems. He also claims that the nature of proof in mathematics is not essentially different from the method of other sciences and methods of arguing in other situations. In [6] a comprehensive theory of knowledge is presented encompassing mathematics.

Many of the points made by the above authors are made because of the emphasis put on the practice of mathematicians, and in particular their experiences. Talking about mathematical experience rather than mathematical reality one wants to emphasize the human aspect of mathematics. The same emphasis also applies to the analysis of proofs. One does not need to reject the presence of objective, mind-independent aspects of mathematics to claim that needs, peculiarities, and limitations of human beings are indispensable for any account of mathematical proofs. They must explain the matter, so some sort of psychologism seems to be inevitable. (See Krajewski [23].) Incidentally, this is another example of an obvious property that is often ignored by those who look for completely objective description, relations between essences, etc.

A much stronger claim to the effect that mathematics is a human activity and nothing more has been made by Rotman. He is close to the view of mathematics as consisting of social constructions

(David Bloor initiated the whole school of sociological account of mathematics; see Ernest [11]). He is, however, watching the behavior of mathematicians in a very penetrating way. In [34] Rotman introduced "a semiotic of mathematics" and pursued the issue further in [35] and [36]. What mathematicians do is described as "thinking and scribbling" performed in order to address other mathematicians. Each mathematician is analyzed into three levels: a mathematical disembodied Subject manipulating signs, above it the real Person with a body and history, telling a metanarrative, and below it a skeletal Agent doing calculations and constructions, also infinite ones, in an imaginary world. A proof is seen as a thought-experiment, and mathematical assertions become predictions about the Subject's encounters with signs.

Let me also mention some other works important for the new philosophical approach. George Polya and his work [29] on non-deductive arguments in mathematics was as an important source, Thomas Tymoczko's influential anthology [40] has served as a reference, Reuben Hersh's anthology [19] gathered together many non-standard approaches to mathematics. In another vein, the book by Stanislas Dehaene [8] on our in-born protomathematical abilities added the neuronal aspect, and the book Lakoff and Núñez [27] emphasized further the fact that our mind is embodied and all the time we use metaphors relating to the physical world.

All varieties of the new, maverick, approach to the philosophy of mathematics share several points. First, the rejection of the Euclidean myth, according to which mathematics is fully objective, completely universal, and absolutely certain. Secondly, a most concentrated attack has been on the idea of the unification of mathematics within one theory, especially on any form of foundationalism, in particular the dominant proposal to have a version of ZF set theory as the foundation. Thirdly and more generally, any imposition of philosophically motivated standards on mathematical activity is rejected. The genuine practice of research mathematicians is declared to be the starting point. This can be expressed, using the term of Penelope Maddy (who, however, wrote as a foundationalist rather than a "maverick"), as "mathematics first", against the traditional "philosophy first" (*philosophia prima*) and the modern "science first." Among the main ingredients of practice is the mathematician's proof.

3. Proofs as They Really Are

In real mathematics problems are proposed and solutions are sought. At the beginning of research for proofs there are problems, not axioms. The work of axiomatizing various domains is also an example of a problem: deciding if given axioms are sufficient for proving a statement is just one more possible math problem. Below, some major features of real life proofs are listed. The proofs must be convincing, understandable, explanatory. (Cf. Hersh [17]: "Proving is convincing and explaining.") Moreover, proofs are meant as valid, final, but at the same time they contain gaps and are revisable.

3.1. Convincing

Most often proofs refer to neither axioms nor other first principles. Instead – as emphasized by Lakatos, Hersh and others – they refer to established mathematics. Whatever is used must be acceptable to appropriate experts. Proofs are presented in the way that makes them understandable to experts. (Textbook proofs for students are often more detailed, but they are fundamentally similar, only a more limited expertise is assumed.) The aim of a proof in a research paper is to convince experts: this category varies according to the context – it can mean all professional mathematicians or, at the other end of the spectrum, a handful of colleagues involved in researching the same topic. In each case a broad corpus of established mathematical results is assumed as given, its validity is not questioned. Of course, mistakes happen. They are, however, sooner or later identified and eliminated. A subtler situation than a simple mistake can occur: sometimes a new understanding of concepts emerges and previous results are rejected or limited to special cases. This was well illustrated by Lakatos who used the Euler formula for polyhedra. Another well-known example, also considered by Lakatos, among many others, is provided by Cauchy's theorem on the continuity of the limit of a converging sequence of continuous functions. Now it is considered a mistake, because uniform convergence must be demanded rather than the weaker pointwise convergence. There exist, however, analyses indicating the correctness of Cauchy's theorem if instead of the current concept of convergence or of the continuum another one is assumed, presumably one closer to Cauchy's original understanding. A perfect example is provided by Robinson's nonstandard analysis: pointwise convergence on standard and nonstandard numbers is sufficient for Cauchy's theorem.

3.2. Understandable

Another psychological property is often assumed by mathematicians: a proof must be understandable. For a human mathematician (are there any other?) one of the most convincing methods of proof is by producing appropriate pictures. This usually enables immediate understanding. Sometimes the picture itself constitutes the proof. Many pictorial proofs of the Pythagorean theorem serve as examples. This sort of proof is possible for many finite configurations, claims Giaquinto [13]; and Brown [2] says that perhaps also for some infinite ones. More than that, often a picture accompanies the invention of a proof in the mathematical "kitchen", to use Hersh's term, even though it rarely finds its way to the official presentation. Even if the matter is not geometric some visual arrangements, mental pictures – imprecise, hazy, messy, often moving, difficult to describe – seem to be common. They help us understand the situation. They are presented to other workers in the kitchen, to help make the point, to convince and induce understanding.

Even when no picture is associated with the proof, to be understandable the proof must be surveyable. Its structure should be graspable. And, preferably, one must be able to tell what is its point. While there are proofs which are not understandable, for example consisting only of calculations, they are seen as less satisfying. And anyway, the discovery of such a proof is usually guided by some understanding. Using Rotman's terms, it is important to be able to have a metanarrative explaining the essence of the narrative that constitutes the proof. This leads us to the next point.

3.3. Explanatory

One of the main features of proof is that it must explain the concepts involved, relations between them, and show not just the truth of conclusion but also *why* the conclusion is true. Often proofs are not providing sufficient explanation, for instance, if the crucial part consists of a calculation and no picture or idea can be indicated as a clarification of the formal manipulations. In such cases a deeper understanding of the proof is sought or other proofs are welcomed so that explanation can emerge. And actually, very often new proofs are sought to explain the aspects of the situation that seem still hidden. Let me mention an example from my own practice. A long time ago I formulated a conjecture (to the effect that a recursively saturated model of Peano arithmetic admits a full satisfaction class – the strict meaning of the terms is not important here) that was soon demonstrated in collaboration with two colleagues and published in Kotlarski, Krajewski and Lachlan [22]. The proof was rather indirect, using a proof theoretic technique. Many years later, long after I had stopped working in this area, Enayat and Visser [9] formulated another proof, much more natural, since it uses only model theoretic constructions. And recently, in 2020, James Schmerl, in the yet unpublished paper "Kernels, Truth and Satisfaction," took the model theoretic proof, and showed that if "stripped to its essentials," it can be expressed as a special property (the existence of a kernel) of certain directed graphs. Thus the technical problem in the proof was reduced to graph theory. The specific logical notions of satisfaction, models, etc. were invoked only as an application of an abstract graph theorem.

Even this modest example illustrates a general point: it is accepted and common to look for a proof by taking advantage of other branches of mathematics than the one in which the problem is formulated. A famous example is provided by Fermat's Last Theorem. Also merging methods and concepts of various branches is seen as valuable, for example probabilistic methods are used in various ways even if probability was not mentioned in the initial problem. New branches were created when similarities of constructions in different parts of mathematics were noticed and properly defined. Or, as a well-known saying goes, good mathematicians perceive analogies, and the best see analogies between analogies. Category theory is a good example.

It is also important to remember that there exist tentative proofs or proofs produced by doubtful methods, for example by analogy. A famous example is provided by Euler's calculations of some infinite sums. He used infinite polynomials as if they had properties similar to the finite cases. In this way he calculated the sum of the series of the reciprocals of the squares of natural numbers as equal to $\pi^2/6$. (See Polya [29, p. 20], or, for example, Putnam [30].) Of course, Euler was aware that his proof was not certain, but when he calculated the initial segment of the series and found it coincide with the proposed number, up to some decimal position, he was convinced that the result was true and the proof fundamentally correct. Later he found a more standard proof.

All the above examples indicate how natural and desirable it is for mathematicians to use unanticipated methods. In other words, proofs can be very far from being pure. Rather, anything is accepted as long as it leads to the aim of deciding the problem one way or another. The idea advocated by logicians that there is an established framework, language, axioms, and proofs are supposed to be conducted within the framework, is simply not true in living mathematics. On the other hand, there is an attractive element to this idea, and actually finding a pure proof of a major theorem established by extrinsic methods is seen as a valuable achievement. To introduce some "purity" one can also formulate a comprehensive theory in which all the methods used to solve the problem are expressible. One can also try to reconstruct the whole proof in set-theoretical language. Such moves are, however, alien to an overwhelming majority of mathematicians. And even if the proof can be reconstructed, it can no more be as convincing, understandable, explanatory as the original argument. I believe that the explanatory power is felt as the single most important feature of proof.

3.4. Revisable

The above-mentioned two examples, Euler's formula for polyhedra and Cauchy's theorem on continuity of the limit of continuous functions, show that proofs are revisable. This is not something mathematicians usually accept. When is a proof seen as good, proper, correct, worth its name? To quote Epstein [10, p. 137] proofs "are meant to be valid." That is to say, it is impossible for the conclusion to be false if the assumptions are true. The proof is supposed to show that something is a fact. Yet new evidence may emerge and the finality of the proof might turn out to be illusory. This possibility is emphasized by all champions of the maverick philosophy of mathematics. How is this possible?

One reason for the collapse of a proof is due to the possibility of changes in our understanding of the concepts used in a proof (cf. the concept of polyhedron). Another reason is due to changes in the standards of rigor (cf. Euler's calculation of the sum of the series of the reciprocals of the squares of natural numbers). Yet another reason is due to the chance of errors that keep popping up. While, as mentioned above, it is generally believed that errors can be ultimately overcome, the more complex the arguments the more probable are either mistakes in proofs or omissions that can be threatening. Some important examples have appeared rather recently, for example enormously long proofs, like the classification of all finite groups that has been achieved by a long collective process involving many mathematicians. There were leaders of the effort, but it seems that nobody has checked the whole proof. (See, for example, Byers [3].) Still it is believed that the job has been done. It is not impossible, though, that something has been overlooked.

Another important kind of example emerged when computers began to be used in mathematics. There exist proofs partly executed by computers. The four color theorem is the best-known example. (See Tymoczko [39]; it was the first philosophical analysis of computer-assisted proofs.) The possibility of error contained in the hardware used is a new source of uncertainty. Yet, repeating the proof on other machines very significantly reduces the chance error. It is probable that the chance human proofs contain errors is higher.

In addition to computer-assisted proofs there are probabilistic proofs. Using it one can prove that a very large number is prime but the proof procedure uses several random moves and is so conceived that it gives the result (that the given number is prime) only with a very high probability. If the chance of error is less than $1/2^{100}$ we can be pretty sure that the result is correct. (See, for example, Rav [32] for more details and references to the papers, from the 1970s, by Michael Rabin and by Robert Solovay and Volken Strassen.)

This last example gives a proof that there are *bona fide* mathematical proofs that lead to conclusions that are not certain. The claim of "the mavericks" is that all proofs share this characteristic. This applies even to most formal ones. As indicated by Cellucci and also by Friend [12, p. 207] even formalized proofs can have "external gaps". These are gaps residing in the external context of proof, specifically in the justification for an axiom or rule of inference. We take it for granted because we assume a standard interpretation. Yet a non-standard interpretation can appear, even of a logical symbol or of a basic concept like that of a set. Then some of the obvious properties may no longer be true. Think of the law of excluded middle which is rejected by constructivists or of the concept of set as defined by a set theory other than ZFC.

4. Conclusion

There is a whole spectrum of the views on the nature of mathematical proofs. An extreme position was expressed by Hardy: there is no such thing as a proof, "we can, in the last analysis, do nothing but point," so there are only rhetorical "devices to stimulate the imagination of the pupils" [14, p. 18]. The other extreme is expressed by Hilbert's Thesis: real proofs are abbreviations and approximations of the ideal formal proofs. Hersh wrote that the belief in the Thesis "is an act of faith" [17, p. 391]. Logicians tend to believe it; their evidence is inductive: so much has been formalized that it seems that we can never encounter insurmountable obstacles if we try hard enough. The point illustrated by the considerations contained in this paper is that even if this is the case and in principle we can convert each proof into a formal one, this is not really significant. The most important features of real proofs – their being convincing, understandable, explanatory – are lost in the process. And the reasons for revisability are not present within the formal proof. The maverick philosophy of mathematics has succeeded in exhibiting the whole range of problems related to

Hilbert's Thesis. The debate on the possibility and significance of formalizability of proofs continues.

References

1. Asprey, W., and P. Kitcher (eds.). *History and philosophy of modern mathematics*, Minneapolis: University of Minnesota Press, 1988.
2. Brown, J. R. *Philosophy of Mathematics. A Contemporary Introduction to the World of Proofs and Pictures*, New York and London: Routledge, 1999, 2008^2.
3. Byers, W. *How Mathematicians Think; Using Ambiguity, Contradiction, and Paradox to Create Mathematics*, Princeton: Princeton University Press, 2007.
4. Cellucci, C. Why Proof? What is a Proof? In G. Corsi and R. Lupacchini (eds.), *Deduction, Computation, Experiment. Exploring the Effectiveness of Proof*, Berlin: Springer-Verlag, 2008, pp. 1-27.
5. Cellucci, Carlo. *Rethinking logic: Logic in relation to mathematics, evolution, and method*, Dordrecht: Springer 2013.
6. Cellucci, C. *Rethinking knowledge: The heuristic view*, Dordrecht: Springer, 2017.
7. Davis, P. J., and R. Hersh. *The Mathematical Experience*, Boston, Basel, Berlin: Birkhäuser, 1981, 1995^2.
8. Dehaene, S. *The Number Sense: How the Mind Creates Mathematics*, Oxford: Oxford University Press, 1997.
9. Enayat, A, and A. Visser. New constructions of satisfaction classes. Unifying the philosophy of truth, pp. 321-335, *Logic, Epistemology, and the Unity of Science* 36, Dordrecht: Springer, 2015.
10. Epstein, R. L. *Five Ways of Saying "Therefore" Arguments, Proofs, Conditionals, Cause and Effect, Explanations*, Belmont, CA: Wadsworth/Thomson Learning, 2002.
11. Ernest, P. *Social Constructivism as a Philosophy of Mathematics*, Albany, NY: SUNY Press, 1998.
12. Friend, M. *Pluralism in Mathematics: A New Position in Philosophy of Mathematics*, Dodrecht: Springer, 2014.

13. Giaquinto, M. Visualising in Mathematics, In P. Mancosu (ed.), *The Philosophy of Mathematical Practice*, Oxford: Oxford University Press, 2008, pp. 22-42.
14. Hardy, G. H. Mathematical Proof, *Mind* 38 (1928), pp. 1-25.
15. Hersh, R. Some proposals for reviving the philosophy of mathematics, *Advances in Mathematics* 31, 31-50, reprinted in T. Tymoczko (ed.), *New Directions in the Philosophy of Mathematics, An Antology*, Basel: Birkhäuser, 1986.
16. Hersh, R. Math Has a Front and a Back, *Eureka* 1988, and *Synthese* 88 (2), 1991.
17. Hersh, R. Proving is convincing and explaining, *Educational Studies in Mathematics* 24, 1993, pp. 389-399.
18. Hersh, R. *What Is Mathematics, Really?* Oxford: Oxford University Press, 1997.
19. Hersh, R. (ed.). *18 Unconventional Essays on the Nature of Mathematics*, New York: Springer, 2006.
20. Hersh, R. *Experiencing mathematics: What do we do, when we do mathematics?* Providence: American Mathematical Society, 2014.
21. Kitcher, P. *The Nature of Mathematical Knowledge*, Oxford: Oxford University Press, 1985.
22. Kotlarski, H., S. Krajewski, and A. H. Lachlan. Construction of satisfaction classes for nonstandard models, *Canadian Math. Bulletin* 24, 1981, pp. 283-293.
23. Krajewski, S. Remarks on Mathematical Explanation, In B. Brozek, M. Heller and M. Hohol (eds.), *The Concept of Explanation*, Kraków: Copernicus Center Press, 2015, pp. 89-104.
24. Krajewski, S. On Suprasubjective Existence in Mathematics, *Studia semiotyczne* XXXII (No 2), 2018, pp. 75-86.
25. Lakatos, I. *Proofs and refutations; The logic of mathematical discovery*, edited by J. Worral and E. Zahar, Cambridge: Cambridge University Press 1976, 2015^2. (Original papers in *British Journal for the Philosophy of Science* 1963-64.)
26. Lakatos, I. A renaissance of empiricism in the recent philosophy of mathematics? In I. Lakatos, *Philosophical Papers*, vol. 2, edited by J. Worrall and G. Curie, Cambridge: Cambridge University Press, 1978, pp. 24-42.

27. Lakoff, G., and R. E. Núñez. *Where Mathematics Comes From. How the Embodied Mind Brings Mathematics into Being*, New York: Basic Books, 2000.
28. Mancosu, P (ed.). *The Philosophy of Mathematical Practice*, Oxford: Oxford University Press, 2008.
29. Polya, G. *Mathematics and Plausible Reasoning*, vol I, Princeton: Princeton University Press, 1973.
30. Putnam, H. What is Mathematical Truth? In *Philosophical Papers*, vol. 1, *Mathematics, Matter and Method*, Cambridge: Cambridge University Press, 1975, pp. 60-78.
31. Rav, Y. Philosophical Problems of Mathematics in the Light of Evolutionary Epistemology, *Philosophica* 43, No. 1, 1989, pp. 49-78; also Chapter 5 in Hersh, R (ed.). *18 Unconventional Essays on the Nature of Mathematics*, New York: Springer, 2006.
32. Rav, Y. Why Do We Prove Theorems? *Philosophia Mathematica* (3) 7, 1999, pp. 5-41.
33. Rav, Y. The Axiomatic Method in Theory and in Practice, *Logique & Analyse* 202, 2008, pp. 125-147.
34. Rotman, B. Toward a Semiotics of Mathematics, *Semiotica* 72, 1988, pp. 1- 35.
35. Rotman, B. *Ad Infinitum ... the Ghost In Turing's Machine: Taking God Out Of Mathematics And Putting the Body Back In*, Stanford, CA: Stanford University Press, 1993.
36. Rotman, B. *Mathematics as Sign: Writing, Imagining, Counting*, Stanford, CA: Stanford University Press, 2000.
37. Sriraman, B. (ed.). *Humanizing Mathematics and its Philosophy*. Essays Celebrating the 90th Birthday of Reuben Hersh, Basel: Birkhäuser 2017.
38. Schmerl, J. Kernels, Truth and Satisfaction, *to be published*.
39. Tymoczko, T. The four-color problem and its philosophical significance, *The Journal of Philosophy* 76, 1979, pp. 57-83.
40. Tymoczko, T. (ed.), *New Directions in the Philosophy of Mathematics, An Antology*, Basel: Birkhäuser, 1986.
41. Wagner, R. *Making and Breaking Mathematical Sense: Histories and Philosophies of Mathematical Practice*, Princeton: Princeton University Press, 2017.

42. White, L. The Locus of Mathematical Reality: An Anthropological Footnote, *Philosophy of Science* 14, 1947, pp. 289-303, reprinted in Hersh, R (ed.). *18 Unconventional Essays on the Nature of Mathematics*, New York: Springer, 2006, pp. 304-319.

43. Wilder, R. L. The Cultural Basis of Mathematics, in *Proceedings of the International Congress of Mathematicians*, AMS 1952, pp. 258-271.

44. Wilder, R. L. *Mathematics as a cultural system*, Oxford: Pergamon Press, 1981.

Necessity and Determinism in Robert Grosseteste's *De libero arbitrio*

Marcin Trepczyński

University of Warsaw
Krakowskie Przedmieście 3 Street
00-927 Warsaw, Poland
e-mail: m.trepczynski@uw.edu.pl

1. Introduction

The concepts of necessity and determinism belong to those philosophical problems which seem to be "immortal": they are discussed by subsequent generations of thinkers, and it is highly likely that they will keep coming back, inspiring philosophers to reconsider them and formulate new insights. Professor Jan Woleński is one of those philosophers who have made successful attempts at discussing these issues and presenting them as clearly as possible. He has accomplished this task both in the context of the problem of free will [27] and within his analysis concerning the topic of the determination of the past and the future [26]. The latter was conducted as part of a discussion inspired by the book by Marcin Tkaczyk on future contingents [21]. Numerous replies and polemics (e.g. [17], [16], [9]) produced in response to this book, as well as other recently published papers (see e.g. [3], [4], [7], [8], [22]), reveal that this topic is still vivid.

However, it is very important to keep in mind former discussions on those issues, including analyses conducted by ancient and medieval philosophers who have provided foundations for later debates (cf. [12]), at least for two reasons. First, we should not neglect the historical approach and instead present the development of ideas and problems in their historical context in order to properly understand the current debates, which are more or less shaped by the past. Second, it is important to accumulate knowledge, including various results obtained by classic authors, since they provide very interesting and inspiring approaches, as was, for instance, the case with Jan Łukasiewicz's proposals concerning the determination of the future, especially his three-valued logic, inspired by Aristotle's *Peri hermeneias* IX with its famous example of the sea battle. What is more, it seems that medieval thinkers were the ones who established the main approaches to the topic of future contingents (cf. [25]). In this light it is worth collecting and analyzing such approaches, both those leading and those less popular, especially since they can still provide us with new solutions. Among recent findings which revealed a worthwhile contribution of medieval thought to such problems were the applications of a version of the principle of the necessity of the past from the works of 12th-century philosophers identified by Wojciech Wciórka, labelled by this author as the "restricted necessity of the past" (RNP), according to which "Every true dictum about the past whose truth does not depend on the future is necessary" [24].

In this article, I would like to present the concept of necessity formulated by Robert Grosseteste (ca. 1168 – 1253) in his work *De libero arbitrio* (*On Free Decision*). I will argue that his approach is an important and inspiring contribution to the discussions on necessity and determinism. What is more, I will show that it can also be useful in contemporary debates, by referring to some issues discussed by Professor Woleński.

Robert Grosseteste was one of the most outstanding thinkers of the Middle Ages, known especially for his treatise *De luce* (*On Light*) in which he claimed that light is the first corporeal form (so every material thing is somehow made of light) and that at the beginning there was a point of light that infinitely multiplied, by auto-diffusion "producing dimensions of space and subsequent beings" [5, p. 104], that is – the world. He is also famous for being one of the first promoters of experimental methods in medieval science. Less known, yet very influential, "penetrating and original" [13, p. 1], is his work *De libero arbitrio* in which he conducts subtle analyses concerning free will or, to be precise: free decision. After Ludwig Baur's edition from 1912, a critical edition of two recensions of this work with English translation has been published in 2017 by Neil Lewis [18], and there are still few studies on this treatise (the most important are: [6], [13], [15]), despite it being very worthy of attention. Thus, it is especially great to see that this work has recently been singled out by

Agnieszka Kijewska, who has perfectly presented Grosseteste's concept of free will and his arguments for the compatibility of freedom of will and God's foreknowledge in the volume *If God exists...*, in a separate chapter [11]. I would like to add to the above-mentioned studies some remarks focusing on the concepts of necessity and determinism. My reflections will be based on Lewis's edition of the later recension, referred to by Baur as recension I, which is the complete one [14, pp. xiv, xix]; however, I will also refer to a crucial point of the earlier recension. When citing this edition, I will give page numbers only; I will quote the Latin version when it is important to show the original wording.

2. Kinds of Determination

As Agnieszka Kijewska [11, pp. 136-137] rightly points out, in *De libero arbitrio*, Robert Grosseteste uses Aristotle's method of analysis (described in *Posterior Analytics*), which requires sorting issues by four questions related to the four things we seek: (1) the fact (*oti*), (2) the reason why (*dioti*), (3) whether it is (*ei esti*), (4) what it is (*ti esti*). Before asking what is (*quid sit*) freedom of decision (chapters 16-19), he decides to analyze whether there is anything like free decision (*an sit*), and to finally (in chapters 20-21) ask about its features, such as what it derives from (*a quo*). Thus, in the introductory sentence he indicates that, first of all, it should be asked: Is there *liberum arbitrium* (hereinafter: LA) at all? And he enumerates possible factors which can "destroy" LA, like: "God's foreknowledge and predestination, the truth of a *dictum* about the future, divination and prophecy, the necessity of fate, grace, and the compulsion to sin that stems from temptation or some kind of force," and "our sinning by means of free decision," as well as "other things that do not come to our mind right now" [p. 109]. These "factors" and their relation to the necessity of future events are analyzed by Robert in subsequent chapters of the first part devoted to the topic of the existence of LA.

When we compare the above-mentioned enumeration and the content of these chapters, we can see that he distinguishes at least five different kinds of determination of future events:
1) someone knows the future (it includes the case of the truth of a *dictum* about the future), so it is determined;
2) God predestines someone, so the effect of predestination is determined;

3) there is fate, so it determines the future;
4) grace makes a deed meritorious totally, so it is thoroughly determined as meritorious;
5) if sin dominates in a human being, it determines that he/she does evil.

Let us note that Grosseteste offers here a really wide range of possible kinds of determination and that they are really diverse. What is more, this list is not necessarily exhaustive, as he is aware that there can be other possible factors that destroy LA. The case marked as (1) represents determination which is not connected with any action or any property of being. It is based on someone's knowledge only. Case (2) refers to God's will and decision which produces a real effect in someone. Case (3) assumes a power ruling all reality in a certain way which cannot be changed. Finally, (4) and (5) refer to the theological reality in which grace or sin can force the human being into some state, and it is impossible for him/her to change it by his/her own, natural powers; yet the two situations are different: grace is given by God, whereas sin is in the human being, and grace not only causes the human being to be able to do good, as it is possible without grace, but it also makes deeds meritorious – thus, it is not just a simple opposition to evil done under the influence of sin.

The most extensive analyses provided by Grosseteste concern the first kind of determination (chapters 1 – 8). Next, Robert briefly deals with: predestination (chapter 9), grace (chapter 10), fate (chapter 11) and sin (chapters 12 – 13). It proves that the first kind of determination, marked above as (1), seems to him to be the most problematic one. And, to anticipate further presentation, we can say that it is with (1) that the most important considerations on necessity are connected. What is more, (1) significantly differs from other kinds of determination. The latter are related to mechanisms to which our actions are subjected (effects of fate, grace, sin) or with God's acts of will (decision to predestine someone). They are supposed to (directly or indirectly) determine some events. It means that there is a kind of causal connection between determinants and determined events; however, this connection may be non-physical. It can be an influence of a supernatural character, as in the case of grace or sin. Therefore, we are speaking here about a broad understanding of cause – it could be an event such as God's decision or human sin, and it could produce another event as a result. On the contrary, in (1) there is no connection based on which a certain event would be a result, and in this way would be determined. It is only assumed that God knows future events or, in a generalized version, that the value of the sentences about future events is set, and that they can be either true or false. Hence, if there is any causal connection, we can only admit that such knowledge or logical value is a consequence of an event.

At this stage, one should note that Grosseteste, within the list discussed above, does not explicitly distinguish cases of physical determinism: neither causal, including anticipants of modern mechanisms, nor teleological ones. It may be bothering. Although such ideas were not popular in the Middle Ages, they were expressed to some extent in ancient philosophy, especially by the ancient atomists, and it seems that a thinker as great and as well-educated as Robert, who quotes Cicero already in the first chapter of *De libero arbitrio*, should be familiar with them. However, it could be argued that in chapter 11, where Grosseteste discusses the topic of fate, he indeed refers to physical determinism. After his presentation of Boethius's concept of fate which relies on God's providence as its consequence, he invokes Cicero's definition of fate (cf. pp. 196-197). According to this ancient philosopher, fate is an order and a series of causes where one cause generates another one, and which flows from God who is an eternal truth and an eternal cause. Cicero argues that this meaning of fate should not be connected with superstition, but it is understood in a "natural sense" (*physice*). Furthermore, Robert also considers another, more "usual," understanding of fate as "the necessity of all lower things that stems from the ordering and turning of the celestial bodies," according to which "clearly everything would happen of necessity and nothing from freedom of decision" [pp. 198-199]. However, we should point out that when referring to Cicero, Robert is not concerned with the purely materialistic view, neither in the version represented by ancient atomists, nor in the teleological perspective of the Stoics, similarly to Marcus Tullius, who criticized both Epicureans and the Stoics for their materialism [19, p. 129]. Robert only considers a possibility that according to God's will the world is ordered by causal connections, and he does not limit them to the states of material objects. Thus, "physics" or "nature" is understood here in a wide sense, not restricted to the material world. What is crucial, Robert argues that some of the events are contingent and thus there is space to include LA. Hence, he indeed avoids considering the causal order in a materialistic manner.

There could be, of course, several reasons for him to exclude this option. He could have treated it as not a serious option. But, at the same time, he took into account an idea such as fate to discuss it and show that it can be understood "seriously," as in the case of Boethius or Cicero. He also could have assumed that if we accept physical determinism, there is no place for LA, so there is nothing to discuss. However, Epicurus's approach revealed that it is possible to combine the concepts of atomism and physical determinism with free will. Finally, one could point out that he intended his treatise to be a theological one, so if he assumed that God created the world and human beings as free creatures with rational souls (cf. pp. 200-201), we should exclude physical determinism. But, if so, he should not be questioning the possibility of free actions and free decisions at all. Finally, we could claim that, following Aristotle, he could have taken it for granted that within the chain of causes we should include acts of the substances which have souls as their principle of movement, hence: independent from external principles, so he found it useless to discuss if souls could be subjected to physical-material determinants. However, it seems that the question about his reasons for not putting this kind of determinism on the above-mentioned list remains open.

In any case, if we want to make use of the list of the kinds of determination presented by Grosseteste, we can supplement it by adding causal determinism understood in a materialistic manner as one of the options of "physical" determinism labelled as fate, together with a materialistic version of teleological determinism, as viewed by the Stoics.

Finally, let us note that some of the options from Grosseteste's list, understood according to his interpretation – are not mutually exclusive. For instance, determination based on foreknowledge, so (1), is compatible with all of the other options. Similarly in the case of determination based on predestination, so (2), if we assume that fate, so (3), is understood – following Grosseteste's view – as compatible with or even as a result of God's will. It also seems that (3) may be compatible with (4) and with (5), for two reasons: (1) the causal arrangement of events, that is – fate, does not interfere with the moral, theological or supernatural qualification of certain deeds, (2) even if a deed is causally determined, it can still be confirmed by the consent of the person performing such a deed.

3. Kinds of Necessity

It is interesting that Grosseteste uses the words "necessity" and "necessary" in many contexts. In chapter 1, when he analyzes the central syllogism:

> Everything known by God is or was or will be; a is known by God (let a be a future contingent); so a is or was or will be [p. 111],

he speaks about the necessity of sentences belonging to this syllogism. First, he assumes that both premises are necessary ("utraque praemissarum est necessaria" [p. 110]). Then he states that if the premises are necessary, then the conclusion is not just true, but also necessary ("conclusio non solum vera, sed etiam necessaria" [p. 110]). This juxtaposition of the truth and the necessity of the sentence shows that he speaks about the necessity of sentences in terms of modal logic.

Moreover, he formulates there an interesting rule of a modal logic which he applies in his considerations: "Ex necessariis enim non sequitur nisi necessarium" [p. 110]. "Non sequitur nisi" means that it is necessary that one follows from another. This means that if Y follows from X, and X is necessary, then it is necessary that Y is necessary. We could put it as follows:

$$(\Box X \to Y) \to \Box (\Box X \to \Box Y)$$

This rule tells us that in a syllogism which is well-constructed, which guarantees the necessity of the inference, necessity of the premises is transferred to the conclusion. Thus in the literature it is referred to as "Transfer of Necessity Principle" (cf. [28]). We should stress that this rule or principle does not have a metaphysical/ontological character, but a logical one. It is clear especially in the case of Grosseteste who, in the cited passage, speaks about the logical inference and the transfer of necessity from premises to a conclusion.

Incidentally, one could doubt whether Robert is actually referring to the necessity of the inference itself, which in the example given above was assumed on the basis of the phrase "non sequitur nisi." But the answer is definitely yes, as he states explicitly that when each of the premises is necessary, the inference is necessary ("patet etiam quod consecutio est necessaria" [p. 110]). It is confirmed by other examples as well. For instance, in a discussion conducted further in the text (chapter 4), we find the phrase: "possem necessario inferre: Socrates est; ergo Socrates est albus" [p. 126]. It means that it is natural to him that, in the case of a well-constructed syllogism, the inference is necessary.

However, in chapter 2, he refers to another kind of necessity which can be considered a metaphysical one. There, he argues that God cognizes singular events (*singularia*). In one of the discussed arguments we read: "cum sit singularium creator, de necessitate cognoscit ipsa" [p. 116]. In this case, necessity is not transferred from another sentence on the basis of a syllogism. The necessity of the fact that God cognizes singular things is a consequence of the fact that God had created them, so we are dealing here with a metaphysical entanglement concerning a rational substance (in this case God) which intentionally creates something: when x intentionally creates y, x cognizes y. It simply results from a necessity which concerns beings.

Robert used this approach after his presentation of a list of arguments concerning the necessity of God's knowledge, concluded by a reflection that "what is contingent does indeed follow from things that are necessary" [p. 129]. At the same time, he pointed out that such a conclusion "goes against the art" [p. 129] in light of the syllogism presented above. However, he explained that "it seems that not only is it possible for what is contingent to follow from things that are necessary, but that this also is necessary"; referring to arguments of such authors as Augustine, Boethius, Seneca (addressing "forms, which Plato calls Ideas") and Anselm of Canterbury, he shows that there is a relationship between necessary reasons (*rationes*), e.g. in God's mind, which are "eternal, stable and unchangeable," and things that are "temporal, changeable, corruptible, and contingent," such that these contingent things "flow" from those necessary reasons [p. 131]. It shows that Robert adopted here a metaphysical approach and started considering properties of different kinds of beings. In this view, it is crucial whether a being *itself* is metaphysically necessary or contingent.

On this basis, he formulates his main solution concerning the concept of necessity, which is very similar in both recensions; however, in the earlier one it is preceded by direct references to the distinctions proposed by Boethius and by Anselm of Canterbury (cf. [11, pp. 139-140]). First, after Anselm, he distinguishes:

1) precedent necessity (*necessitas praecedens*), which is – as Grosseteste puts it – "a cause of a thing's existence and forces the thing to exist";
2) sequent necessity (*necessitas sequens*), which is not such a cause and "does not force a thing to exist," "which produces nothing" and just "seems to destroy alternatives" (as it refers to what is contingent), e.g. "while I am sitting, it is necessary that I am sitting" [pp. 22-25] – in this case the alternative "I am sitting or I am not sitting" is destroyed.
Next, he reformulates the two kinds of necessity and calls them:
1) necessity from which "only what is necessary follows";
2) necessity from which "what is contingent seems to (*videtur*) follow."

Then, he refers to Boethius and says that, in his opinion, these kinds of necessities were called by Boethius: simple necessity (*necessitas simplex*) and necessity of condition (*necessitas condicionis*) [pp. 24-25], where an example of the first one is that it is necessary that all human beings are mortal, and an example of the latter is that if we see someone walking it is necessary that he/she is walking (Boethius, *De consolatione philosophiae*, V.6; cf. [2, p. 148]).[1] Robert underlines that there are some who mean something else by necessity of condition, namely necessity of consecution of the consequent from the antecedent, but in fact ("in more depth") Boethius refers there to what Robert calls sequent necessity.

Now, in both recensions Grosseteste presents, in a very similar way (so I will now quote the later recension again), a crucial division, according to which something may be:
1) necessary "unqualifiedly" (*simpliciter*), which means that "it has no capacity (*posse*) at all for its opposite, either with or without a beginning," e.g. "that two and three are five";
2) necessary in a way that "it has no capacity for its opposite in respect of the past, present, or future, yet without a beginning there was a capacity for it and a capacity for its opposite" [pp. 134-135].

What may be most puzzling here is the expression "without a beginning." We should note that it refers to a sort of pre-temporal ("before time") or even atemporal perspective, in which we abstract from time and from the fact that a thing started to exist (had a beginning) or even from the fact that anything started to exist (as it can refer to the whole world). This idea seems to be connected with the Augustinian and Boethian concept of eternity, according to which

God is not subject to change and time, and sees everything at once (which explains why Robert also uses the double expression: "from eternity and without a beginning"). Both authors are often quoted by him, also in this respect. In chapter 2, Grosseteste refers to Book V of Boethius's *De consolatione* (in which "the last Roman" formulated his famous definition of eternity: "aeternitas igitur est interminabilis vitae tota simul et perfecta possessio") and quotes some passages, including the one about God's "single mental glance" and the one according to which "God sees at present the future things" [pp. 114-115]. And in chapter 3, among many passages from Augustine, he quotes the one from *Confessiones* (XII, 15) about God's will being unchangeable, in which the bishop of Hippo states that "everything changeable is not eternal, but God is eternal" [pp. 124-125]. It means that God does not "live" in time and sees all the past, present and future events at once. According to Robert's above-mentioned expression, it also includes a perspective in which such events are still not actualized, they are still possible.

In this context, Grosseteste shows that there are such propositions or *dicta* that if their truth is established, it cannot cease (their truth will not have "non-being after being"), so they do not have the capacity for their opposites (as "they cannot be altered from being true to being false"), but if we abstract from the fact that they are already true and adopt an eternal, atemporal perspective, they still have such a capacity. As Robert explains, the *dictum* "that Antichrist will be going to exist" is true in respect of the past, the present and the future, as it has no capacity for its opposite, but "without a beginning" it does have such a capacity. So for such *dicta*, as well as for some true *dicta* about the future (e.g. given in prophecy), "from eternity and without a beginning" there is a capacity to have been true and a capacity to have been false. And, in this sense, things that such *dicta* are about are contingent (cf. pp. 134-135). So if "God knows *a*" is true, then it cannot become false, and in this way it is necessary. The same situation arises when we consider the conclusion of a syllogism drawn from such true (so: necessary) propositions. However, from the perspective "without a beginning," the opposite is possible, and in this way *a* is contingent.

On this basis Robert discerns:
1) necessity in the sense that the truth of the sentence cannot cease;
2) necessity of existence.

In this light, he claims that "Antichrist necessarily is going to exist" has two interpretations, as the necessity can apply to:
1) "the futurity attributed to Antichrist, and in this sense it is true and follows by syllogistic inference from premises that are necessary in the same sense";
2) "the existence of Antichrist, which is future, so that the sense is 'Antichrist will have in the future existence of necessity', and in this sense it is false and does not follow from any premises that are either true or necessary, for in the future he will have contingent existence" [pp. 136-137].

So he concludes that in such *dicta* "a certain contingency is combined with a certain necessity." Hence, he discerns three cases:
1) "total necessity" (*omnino necessitas*), as in "two and three are five";
2) "necessity in one respect and contingency in another," as in some true *dicta* about future contingents (or present or past), namely those whose truth cannot cease;
3) "total contingency" (*omnino contingentia*) in *dicta* which are true, but can become false, like "that Socrates is pale" (cf. pp. 136-137).

Finally, when speaking about contingency within the "without a beginning" perspective, Robert refers to the "contingency of things in themselves" (*contingentia rerum in ipsis*) (cf. pp. 138-139). So, in this case, he applies a clearly metaphysical approach.

To conclude, we should note that Grosseteste presented different distinctions in order to defend contingency, even if we accept a necessity following from God's foreknowledge or even omniscience, starting with the theories developed by Anselm of Canterbury and Boethius, and finally offering his own, original solution. Those earlier theories were based on the analysis of implication representing the connection between cause and effect, where if the occurrence of an effect means that its cause is necessary, it constitutes "sequent necessity" or "necessity of condition," which does not exclude the contingency of such a cause. However, Grosseteste decided to present a theory based on a concept of capacity for the opposite, which can be applied to sentences, propositions, *dicta* or things they are about. Their capacity for their opposites can be assessed from two perspectives: the one in which their truth or falsity is already established, and the one called "from eternity and without a beginning." Such a capacity viewed "without a beginning" is combined with the contingency of a thing in itself, which has a metaphysical character, whereas if such an incapacity is considered in respect of past, presence, or future, we can claim that it has a logical or epistemological character, as it is a consequence of an inference or of a cognition.

A juxtaposition of those two criteria (first: having such a capacity or not, second: the perspective adopted to asses it) gives us the three discussed options. It is obvious that Grosseteste excluded a fourth one, according to which a true *dictum* would have a capacity for its opposite in respect of time, but it would not have such a capacity "without a beginning."

4. Usefulness in Contemporary Debates

To illustrate the ways in which Grosseteste's theory can be useful in contemporary debates I will present selected examples, referring to the issues discussed by Professor Woleński.

A very important contribution to the debate concerning determinism is the division between holistic determinism and distributive determinism. The thesis of the former is the following: "The later states of the world are precisely determined by its earlier states." It is "holistic," as it concerns the world as a whole. Whereas the thesis of the latter is restricted to single events (so the world is treated in a distributive way): "For each event z, z is precisely defined by a set of prior conditions" [27, p. 177]. It is clear that the first one excludes free decision. Professor Woleński underlines that its "rule" includes the cases of the acts of (more or less free) decision, so if we consider them as a part of the set of all earlier states of the later states, it may appear to be compatible with the existence of free decision. However, if we take into account that they are also a part of the later state of the world which is precisely determined, then we see that they cannot be free. But it is different in the case of distributive determinism. Here we can just include the acts of free choice as part of the prior conditions. It seems that we do not need to agree that such acts are also events which are precisely determined.

We should note that in Grosseteste's *De libero arbitrio* we can find examples which can illustrate both of these kinds of determinism. Fate, according to the second understanding, which he finds in Cicero, seems to be the case of holistic determinism, as this approach assumes that everything is necessary and the stars have precisely determined the states of the world. And it seems that, as Robert adopts natural causality (though including the acts of free decision), his general view is an example of distributive determinism. However, Grosseteste provides a wider perspective with his list of kinds of determination, which is important if we want to consider a theological perspective. Also in philosophical debates which include such a perspective it is important to take into account such realities as: sin, grace and God's pre-election (predestination) and its relationship with LA. And Grosseteste provides a subtle discussion concerning all these levels of possible determination. I believe that in such debates it is essential to take into consideration as wide a range of kinds of determination as possible and not to restrict it to a specific case of determination. An example of such a general and broader approach with many types of determination (including: prospective, retrospective and functional ones), together with a reminder that we should not neglect this kind of perspective, has been presented for instance by Jacek J. Jadacki [9, p. 84].

Next, in contemporary debates on determinism and necessity it is important to establish relationships between such notions and their precise meanings. For instance, one can say that "A is determined" is equal to "A is necessary," and "A is contingent" is equal to "A is possible and $\neg A$ is possible," and also "$\neg A$ is determined" is equal to "A is impossible", as Professor Woleński proposed (cf. [26, pp. 188-189]). However, it is possible to do it differently, for example by saying that contingency is expressed simply by the proposition "A is possible" or "It is not true that A is determined," and it is a matter of choice. What is more, such choices may be connected with additional assumptions, like the formal equivalence of the *de re* and *de dicto* interpretation (cf. [26, p. 185]), whatever such distinction would mean exactly, bearing in mind that we can find different approaches to this division.

Grosseteste provides a clear theory in which necessity is defined as incapacity for the opposite. It means that if something is necessary it has no capacity (or possibility – *posse*) for its opposite. And something is contingent when it has such a capacity (so: possibility). This category is universal, as it may be applied – as Robert shows – to different perspectives, like: (1) as referred to the past, presence or future, in which the truth/falsity of the *dictum* can be already established, (2) "without a beginning and from eternity." It seems that his theory may be easily translated into the theory of possible worlds in the following way: necessity in (1) is necessity in an actual (or a chosen) world, whereas necessity in (2) is necessity in every possible world. At least for these reasons it is worth referring to Grosseteste's theory when establishing the meaning of these notions in our debates.

Furthermore, Robert's approach may deepen such debates by introducing (with the above-mentioned perspectives) two important dimensions. It can, for instance, shed a new light on the following simple triad of possible relationships between the notions of actuality and determination:

(a) if A is actual, A is determined;
(b) if A is actual, A is contingent;
(c) if A is actual, A is determined or A is contingent.

According to Jan Woleński these represent respectively: radical determinism (RD), radical indeterminism (RI) and together moderate determinism (MD) and moderate indeterminism (MI), which are formally indistinguishable [26, p. 188]. When we do not include those two perspectives, we rather interpret "or" in (c) as a disjunction (so "XOR"): A can be determined or contingent, but it cannot be both determined and contingent. By contrast, the juxtaposition of those two perspectives, offered by Robert, produces an option such that A is actual and A is both determined (according to (1)) and contingent (according to (2)). This approach generates the following matrix of positions, where (1) and (2) indicate the above-mentioned perspectives:

(a)   RD(1)-RD(2);
(a') RD(1)-MD/MI(2);
(a") RD(1)-RI(2);
(b)   RI(1)-RD(2);
(b') RI(1)-MI/MD(2);
(b") RI(1)-RI(2);
(c)   MD/MI(1)-RD(2);
(c') MD/MI(1)-RI(2);
(c") MD/MI(1)-MD/MI(2).

The position that Robert represents is, of course, (a'), as God knows about every state of affairs (in this way they are actual), so for him in (1) their truthfulness is established (and in this way they are determined), but in (2) some of these states of affairs are necessary and some are contingent. According to Grosseteste's approach, (b) – (b″) would mean that God knows nothing about the world. Those options could represent Aristotle's theory where God thinks about himself only or a position that God does not think or know anything, or simply atheism. What is more, to accept (b) – (b″) we would have to admit that not only the sentences about future events would not be determined, but also all the sentences about the past and the present ones. Perhaps, we could label such a position as a kind of radical skepticism. And (c) – (c″) mean that if there is God, his knowledge is partial. Perhaps, there can be other interpretations of the options presented above, and for sure there are many interesting problems arising from such a matrix, but this deserves a separate discussion.

Finally, when talking about God's omniscience, it is very easy to slip into a deterministic perspective, from which it is very difficult to see any solution defending the possibility of free decision. Professor Woleński has rightly pointed out that, in terms of such omniscience, there is no difference between foreknowledge and knowledge, in particular between truths about the past and truths about the future. On this basis he concluded:

This, in turn, means that the set of future contingents in this situation is empty – also because divine knowledge is necessary. The problem for the theologian is that, if the world was created by God, fatalism with reference to human choices seems to be unavoidable [26, p. 193].

Grosseteste's great effort was to show that such fatalism is indeed avoidable. But at the same time, he was aware that debates on this topic are often connected with a pre-supposition concerning incompatibility. He indicated the source of the problem as follows:

> So the fog that surrounds these matters is wholly a product of the fact the contingency of things in themselves seems to be incompatible with their necessity in the divine mind and knowledge [...]. It is also a product of the fact that it is not distinguished how in the same proposition in one respect there is necessity [...] and in another respect contingency [p. 139].

And in my opinion he really did provide a well-constructed theory and argumentation to defend the compatibility of LA and God's omniscience. Therefore, this is another reason to refer to his work anytime we start a debate on these topics.

5. Conclusions

Robert Grosseteste presented a wide range of different kinds of determination of human actions and used the notion of necessity in various contexts, showing the richness of it.

He elaborated an original theory of necessity and contingency based on the concept of capacity for the opposite and two perspectives: (1) being true in respect of the past, present or future and (2) without a beginning and from eternity. This enabled him to explain that determination which follows from the fact that something is true does not exclude contingency, and in consequence to defend the compatibility of God's omniscience and human free decision.

His interpretations and his theory can be useful in contemporary debates. They widen the scope of analyses by adding the "theological" aspects of determination. They provide precise definitions of necessity and contingency, which are a good analytical tool and which can be translated into the concept of possible worlds. Through their two-perspectives approach they enrich the range of options by generating nine kinds of positions on the spectrum of determinism and indeterminism.

It seems that his solution should be taken into account anytime philosophers discuss the topic of determination, particularly determination of the past (or present or future), or of the compatibility of God's omniscience and human free decision. In many situations it may turn out that contemporary arguments concerning these topics are weak or insufficient in confrontation with Grosseteste's solution.

Acknowledgments

This article was supported by the National Science Centre, Poland, under grant agreement No. 2015/18/E/HS1/00153.

References

1. Andreoletti, G. Fatalism and Future Contingents, *Analytic Philosophy* 60 (3), 2019, pp. 245-258.

2. Boethius. *Consolation of Philosophy*, transl. by J. C. Relihan, Indianapolis and Cambridge: Hackett Publishing, 2001.
3. Ciuni, R., and C. Proietti. The Abundance of the Future: A Paraconsistent Approach to Future Contingents, *Logic and Logical Philosophy* 22, 2013, pp. 21-43.
4. Ciuni, R., and C. Proietti. Future Contingents, Supervaluations, and Relative Truth, In L. Bellotti, L. Gili, E. Moriconi, and G. Turbanti (eds.), *Third Pisa Colloquium in Logic, Language and Epistemology: Essays in Honour of Mauro Mariani and Carlo Marletti*, Pisa: Edizioni ETS, 2019, pp. 69-88.
5. Crombie, A. C. *Robert Grosseteste and the Origins of Experimental Science 1100-1700*, Oxford: Clarendon Press, 1953.
6. Dawson, J. G. Necessity and Contingency in the *De libero arbitrio* of Grosseteste, In *La filosofia della natura nel Medioevo: Atti del terzo Congresso internazionale di filosofia Medioevale, Passo della Mendola (Trento) 31 agosto-5 settembre 1964*, Milan: Società editrice Vita e pensiero, 1966, pp. 357-362.
7. Di Nucci, E. Knowing Future Contingents, *Logos and Episteme* 3 (1), 2012, pp. 43-50.
8. Iwanicki, M. Divine Foreknowledge and Human Freedom: N. Pike contra A. Plantinga, In A. P. Stefańczyk and R. Majeran (eds.), *If God Exists... Human Freedom and Theistic Hypothesis: Studies and Essays*, Lublin: Towarzystwo Naukowe KUL, 2019, pp. 425-462.
9. Jadacki, J. J. Causal and Functional Determination vs. Foreknowledge about the Future, *Roczniki Filozoficzne* 66 (4), 2018, pp. 81-98.
10. Jadacki, J. J. O antynomii zdarzeń przyszłych, *Przegląd Filozoficzny* 3 (99), 2016, pp. 311-329.
11. Kijewska, A. Robert Grosseteste and His *De Libero Arbitrio*, In A. P. Stefańczyk and R. Majeran (eds.), *If God Exists... Human Freedom and Theistic Hypothesis: Studies and Essays*, Lublin: Towarzystwo Naukowe KUL, 2019, pp. 133-149.
12. Knuuttila, S. Medieval Approaches to Future Contingents, *Roczniki Filozoficzne* 66 (4), 2018, pp. 99-114.
13. Lewis, N. The First Recension of Robert Grosseteste's *De libero arbitrio*, *Mediaeval Studies* 53, 1991, pp. 1-88.
14. Lewis, N. Introduction, In Robert Grosseteste, *On Free Decision*, N. Lewis (ed.), Auctores Britannici Medii Aevi, Oxford: Oxford University Press, 2017, pp. xiii-cxvi.

15. Lewis, N. *Libertas arbitrii* in Robert Grosseteste's *De libero arbitrio*, In J. Flood, J. R. Ginther, and J. W. Goering (eds.), *Robert Grosseteste and His Intellectual Milieu: New Editions and Studies*, Papers in Mediaeval Studies 24, Toronto: PIMS, 2013.

16. Łukasiewicz, D. Marcin Tkaczyk's Ockhamism, or Whether the Theory of *Contingentia Praeterita* is the only Plausible Solution to the Problem of *Futura Contingentia*, *Roczniki Filozoficzne* 66 (4), 2018, pp. 115-134.

17. Pawl, T. A Reply to "The Antinomy of Future Contingent Events", *Roczniki Filozoficzne* 66 (4), 2018, pp. 149-157.

18. Robert Grosseteste. *On Free Decision*, N. Lewis (ed.), Auctores Britannici Medii Aevi, Oxford: Oxford University Press, 2017.

19. Smith, G. B. *Political Philosophy and the Republican Future: Reconsidering Cicero*, Notre Dame, IN: University of Notre Dame Press, 2018.

20. Tkaczyk, M. The Antinomy of Future Contingent Events, *Roczniki Filozoficzne* 66 (4), 2018, pp. 5-38.

21. Tkaczyk, M. *Futura Contingentia*, Lublin: Wydawnictwo KUL, 2015.

22. Todd, P., and B. Rabern. Future Contingents and the Logic of Temporal Omniscience, *Nous* 2019, pp. 1-26.

23. Wawer, J. *Branching Time and the Semantics of Future Contingents*, PhD dissertation, Kraków, 2016.

24. Wciórka, W. Mitigating the Necessity of the Past in the Second Half of the Twelfth Century: Future-Dependent Predestination, *Vivarium* 58, 2020, pp. 29-64.

25. Wojtysiak, J. Future Contingents, Ockhamism (Retroactivism) and Thomism (Eternalism), *Roczniki Filozoficzne* 66 (4), 2018, pp. 159-182.

26. Woleński, J. Is the Past Determined (Necessary)? *Roczniki Filozoficzne* 66 (4), 2018, pp. 183-195.

27. Woleński, J. Wolność, determinizm, indeterminizm, odpowiedzialność, *Śląskie Studia Historyczno-Teologiczne* 29, 1996, pp. 176-179.

28. Zagzebski, L., Foreknowledge and Free Will, In E. N. Zalta (ed.), *The Stanford Encyclopedia of Philosophy* (Summer 2017 Edition), https://plato.stanford.edu/archives/sum2017/entries/free-will-foreknowledge/.

Notes

1. In the later recension those concepts are used only in the solutions to the other problems, such as free decision and predestination, cf. pp. 184-185: "Accordingly, our reply is that predestination in fact is a necessary cause and has a necessary effect. But it is not an unqualifiedly necessary cause (*non simpliciter*), but conditionally (*condicionaliter*) [necessary], and it has subsequent rather than precedent necessity."

Logical Consequence Operators and Etatism

Wojciech Krysztofiak

The University of Szczecin
Krakowska 71-79 Street
71-017 Szczecin, Poland
e-mail: wojciech.krysztofiak@gmail.com

Each discourse is governed by an inferential mechanism enabling its deductive processing. A peculiar feature of all ideological discourses is that their participants in the processes of developing various narratives form statements banned from different points of view. For example, within religious discourse, atheists utter blasphemous statements from the point of view of followers of specific religions, and theists formulate sentences judged by atheists as insulting human reason. Both sides of the ideological war accuse each other of offending acts, while prohibiting the opposite party from expressing certain sentences classified as blasphemy, offense or hate speech. Even logically correct inference acts are often stigmatized in the ideological exchange by the value of blasphemy or offense, which makes them unacceptable to the parties of the conflict.

The article presents the theory of operators of logical consequence indexed by taboo functions. It will be shown that every discourse in any phase of its development is correlated with a certain logical structure consisting of an open set of discourse sentences, a set of taboo functions and a set of operators of logical consequence indexed by taboo functions. This structure determines the mechanism of deductive processing of sentences produced within a given discourse by its participants. A characteristic feature of these deduction processes is that the same rules of inference are valid in certain narrative contexts of a given discourse and lose their logical validity in other narrative contexts. The presented theory of logical consequence which is a generalization of Tarski's theory, explains the phenomenon of the lability of deduction rules in content processing, in particular within ideological discourses.

On the basis of the presented theory, it is possible to construct idealization models of various developmental stages of the discourse: its etatization, totalitarization, terrorization, de-etatization (liberalization) and its full liberalization phase. The phase of discourse etatization consists in the growth of consequence operators indexed by taboo functions in its logical structure, while the de-etatization phase is an inverse process which culminates in correlating the discourse with a structure comprising exactly one operator of logical consequence called the liberal consequence operator that satisfies the standard conditions for consequence operators specified in Tarski's general axioms. From this point of view, the classical logic, determined by the consequence operator which meets Tarski's conditions, appears as an "oasis of freedom in deduction processes", while the other taboo-indexed operators contribute to the dissemination of penalizing activities of discourse participants. Transformations of various taboo structures of deduction in the course of the historical development of discourse are enabled by penalty functions correlated with corresponding taboo functions. Their mode of action determines, among others, such phenomena as totalitarization and terrorization of discourse.

1. The Phenomenon of Lability of Inference Rules in Discourse Development Actions

Some participants in the discourse recognize the logical correctness of certain inferences, although they assess them as unacceptable at the same time. Here is an example of such inference:

(1) *Jesus Christ is God, therefore Jesus Christ is a cheater or God.*

Some students who have mastered the competence of proving on the basis of classical propositional calculus, state that (i) the presented inference is logically correct and that (ii) the premise is true, and yet (iii) they do not accept the conclusion. However, the same students are able to recognize the correctness and the conclusion of another inference:

(2) *Hitler was the leader of Germany, so Hitler was a bandit or leader of Germany.*

Both inferences fall under the same correct rule of inference of the classical propositional calculus, namely the rule of introducing a disjunction. The presented example shows the lability of inference rules in discourse development actions, which means that in some contexts some discourse participants accept the correctness of inferences carried out in accordance with the correct rules of a given logic, and in other contexts they do not accept the correctness of inferences implemented according to the same rule, although they accept the premises for such unacceptable inferences.

Another manifestation of the lability of inference rules can be observed in relation to the ways of using, for example, Modus Ponens. Some people who efficiently use classical propositional calculus do not want to accept the following inference:

(3) *If the Buddha is God, then Buddhists are stupid. The Buddha is God. So Buddhists are stupid.*

In the case of inference (3), some language users do not want to accept the conclusion due to the rejection of the first premise. In addition, they declare on this basis that all reasoning is logically invalid. However, the same people are willing to accept the logical validity of another inference, even though they recognize the falseness of the second premise:

(4) *If Satan exists, then Satanists are stupid. Satan exists. So Satanists are stupid.*

The above-described facts can be explained by adopting the following two hypotheses:

(1) If the person O conducts inference on the basis of a classical propositional calculus, in which the premises or the conclusion contain an offensive (prohibited, blasphemous, taboo-breaking) sentence, then in the mind of the person O the mechanism blocking the inference is activated, by which (i) O rejects a correctly inferred conclusion on the basis of accepted premise or (ii) O rejects the logical validity of the inference.

(2) If the person O conducts inference on the basis of a classical propositional calculus, in which the premises or the conclusion do not contain a sentence offensive to him, then in the mind of the person O the mechanism preventing the inference is not activated, as a result of which the person O (i) accepts a correctly inferred conclusion from accepted premises and (ii) accepts the logical validity of the inference.

The lability of inference rules consists in that they are judged to be valid in some contexts but invalid in other contexts of the same discourse. This means that the deduction rules acquire their logical validity due to specific properties of the contexts in which they are applied. Such a property is the stigma of being forbidden, offensive or blasphemous in a given context. The comprehension of the inferential context by the participant of the discourse through the stigma of the ban in the inferences presented to him activates a mental mechanism blocking the process of context processing according to a given rule, which in turn triggers the act of rejecting the conclusion regardless of the acceptance or rejection of premises, or triggers the act of assessing the inference as incorrect. If the participant in the discourse does not capture the inferential context through the stigma of the ban in the inference presented to him, then, on the basis of his logical competence, he (she) accepts the derived conclusion or accepts the inference.

Does the presented mechanism blocking deductive processes in the minds of discourse participants have a logical character in the sense that it can be described by a specific structure of deduction? To perform deductive processing of formulas belonging to discourse D, the mind must associate a set D with a specific operator of logical consequence C_i. Let CN be any set of logical consequence operators. The deduction structures are understood as systems of the form: $<D, CN>$. These deduction structures, which are associated with scientific discourses, have the form: $<D, \{C_i\}>$. In this case, the CN is a one-element set. For example, the deduction structure for Peano's arithmetic is a system of the form: $<J\ (PA), \{C_{KL}\}>$, where $J\ (PA)$ is the set of all formulas written in the PA language, and C_{KL} is the operator of the consequence of classical logic.

The hypotheses presented above suggest, however, that the structures of deduction associated by the mind carrying out inference actions within a given discourse in the context of offensive, blasphemous or forbidden sentences are systems with at least two different logical consequence operators, i.e. systems of the shape: $<D, \{C_i, C_k\}>$. The operator C_i is responsible for the deduction processes carried out by the mind within a discourse D in a situation where the mind does not capture the inferential context with the stigma of language taboo. The operator C_k, in turn, cancels the logical validity of inference established by C_i and carried out in contexts with the stigma of the ban (taboo). Metaphorically speaking, the C_k detautologizes some inferences that are tautological from the point of view of C_i.

The described situation can be generalized in such a way that in the deduction structure there are many consequence operators that detautologize some tautological inferences established by other consequence operators. For example, one thing offends a follower of Judaism in statements of a Catholic believer, another thing offends an adherent of Islam in statements of an Old Testament follower, and yet another thing can be a language taboo from the point of view of an atheist Bolshevik in the statements of an Islamist, Catholic or a follower of Judaism. The following reasoning may be, for instance, rejected by some Catholics and fully accepted by Islamists:

(5) *If God is great, he punishes the death of blasphemers. God is great. So God punishes the death of blasphemers.*

Many Catholics (personalists) do not have to recognize (5) as correct reasoning because of its offensive nature. According to (5), God kills people. In turn, some atheists can agree with the Islamist and recognize the logical validity of the presented inference only because it is a substitution of the Modus Ponens scheme.

The theory of operators of logical consequences indexed with taboo functions[1] constructed in the article is a tool that allows to explain the mechanism of detautologizing inferences that are logically valid from the point of view of certain logical consequence operators and at the same time invalid from the point of view of other competing logical consequence operators.

2. Theory of Logical Consequence Operators Indexed with Taboo Functions

The subject of the study of the theory of logical consequence operators indexed with taboo functions is a structure in the form (hereinafter called the structure of deduction with taboo functions): <D, CN, T>, where <D, CN> is the logical structure of deduction of discourse D understood as a set of its formulas, and T is any set of taboo functions. Thus, <D, CN, T> structures are an extension of deduction structures of the shape: <D, CN>. The domain of each taboo function associated with discourse D is exactly one object, which is the set of all formulas of D. Taboo functions can be understood as representations of various institutions of "elm experts" operating within a given discourse.[2] One of the roles of these experts is to control the deduction processes carried out by participants in a given discourse. Within a given discourse, there can be many experts competing with each other or fighting each other, thus designating different operators of logical consequence. Taboo functions and consequence operators indexed with these functions therefore satisfy three general conditions:

A1 $(\forall i)(i \in T \rightarrow i \subset \{D\} \times 2^D)$
A2 $(\forall i)(i \in T \wedge C_i \in CN \rightarrow C_i \subset 2^D \times 2^D)$
A3 $(\forall i, k)[i \in T \wedge k \in T \wedge C_i \in CN \wedge C_k \in CN \rightarrow (i \neq k \equiv C_i \neq C_k)]$

According to A1, each taboo function i maps the set of all discourse D formulas into a subset constituting the language taboo of discourse D according to function i. In turn, according to A2, consequence operators indexed with taboo functions map subsets of set D into subsets of set D. In addition, under A3, the two taboo functions are different when the consequence operators indexed by these functions are also different. This axiom sets the correlation between each taboo function and its corresponding unique logical consequence operator. From the axiom A3, one can conclude that if the set of taboo functions associated with discourse D is one-element, then the set of operators of the consequences CN is also one-element.

T1 $(\forall i, k)(i \in T \wedge k \in T \rightarrow i = k) \rightarrow (\forall i, k)(C_i \in CN \wedge C_k \in CN \rightarrow C_i = C_k)$

Let's adopt the following language conventions:

(i) Variables: i, j, k, l run a set of T taboo functions associated with discourse D in its specific development phase;

(ii) Variables: $C_1, ..., C_i, C_j, C_k$ run a set of consequence operators indexed with functions from the set T;
(iii) Variables: $X, Y, Z, X_1, ..., X_n$ run a power set 2^D;
(iv) Variables: $\alpha, \beta, \gamma, \delta$ run a set of formulas D.
(v) $\aleph$ is a force of countably infinite set and *Card* is a cardinality function.

Other axioms of the constructed theory are as follows:

A4 *($\forall i$)($\forall \alpha$){$i \in T \to [\alpha \in i(D) \equiv \sim(\exists X) \alpha \in C_i(X)$]}*
A5 *($\forall i$)($\forall k$)($\forall X$)[$i \in T \wedge k \in T \to C_i(X - (i(D) \cup k(D)) = C_k(X - (i(D) \cup k(D))$]*
A6 *($\forall i$){$i \in T \to [X \subset i(D) \to C_i(X) \subset C_i(\emptyset)$]}*
A7 *($\forall i, k$) {$i \in T \wedge k \in T \to [i(D) \subset k(D) \to (\forall X)(C_k(X) \subset C_i(X)$]}*
A8 *($\forall i$)($\forall X$)[$i \in T \to X - i(D) \subset C_i(X - i(D))$]*
A9 *($\forall i$)($\forall X$)[$i \in T \to C_i C_i(X) \subset C_i(X)$]*
A10 *($\forall i$)($\forall X$)($\forall Y$){ $i \in T \to [X \subset Y \to C_i(X) \subset C_i(Y)$]}*
A11 *($\forall i$)($\forall \alpha$)($\forall X$){ $i \in T \to [\alpha \in C_i(X) \to (\exists Y)(Y \subset X \wedge Card(Y) < \aleph \wedge \alpha \in C_i(Y)$]},*

Axiom A4 states that if a given formula is banned from the point of view of any taboo function belonging to the class *T* (if it belongs to any language taboo), i.e. belonging to the value of any taboo function, then there is no set of formulas in *D* from which the given formula would be derivable according to the operator of the consequence indexed by a given taboo function. In addition, according to A4, if a formula has the property that there is no set of formulas from which it is inferable according to the operator of the consequence indexed by a given taboo function, then this formula belongs to the set of banned formulas designated by a given taboo function. Hence, axiom A4 expresses a property which can be named the principle of inferential sterility of formulas belonging to any taboo from the point of view of a given taboo function. The same formula, sterile from the point of view of a given taboo function, does not have to be sterile inferentially from the point of view of another taboo function associated with discourse *D*. In light of the axiom A5, two consequence operators indexed with any taboo indexes, acting on any set of formulas disjoint with the set of formulas banned according to one or the other taboo index, return the same set. In other words, any two consequence operators indexed by different taboo functions behave logically the same, acting on sets of formulas not banned from the point of view of the sum of the values of these two taboo functions. A6 expresses that any subset of a given set of banned formulas has the property that the set of formulas derived from it, according to the operator of the consequence indexed by the taboo function that creates a given set of banned formulas, is included in the set of formulas derived according to this operator from the empty set. If the set of consequences of an empty set is an empty set, then no formula is derived from any set of banned formulas. Axiom A7 states that if the set of banned formulas designated by a given taboo function is included in the set of banned formulas designated by the second taboo function (the first taboo function is weaker than the second, stronger taboo function), then the set of formulas derived according to the operator of the consequence indexed by the second taboo function (stronger) from a given set of formulas is contained in a set of formulas derived from the same set of formulas according to the consequence operator indexed by the first taboo function (weaker). In other words, the weaker the taboo function is, the stronger the inferential force of the consequence operator indexed by a given taboo function is, and vice versa, the stronger the taboo function is, the weaker the inferential force of the consequence operator indexed by a given function is. According to A8, any set of formulas minus the formulas belonging to the set of banned formulas, designated by a given taboo function, is included in the set of consequences indexed by this function of a given set of formulas minus banned formulas. In other words, only these formulas are inferable from themselves according to the consequence operator indexed by a given taboo function, which do not belong to the set of banned formulas designated by this taboo function. Other axioms: A9, A10, A11 impose on any consequence operators indexed by taboo functions such properties as: idempotence, monotonicity, and finiteness.

The presented axiom system is a generalization of Tarski's logical consequence theory. If an axiom of the form: (TA) $(\forall i)\ i(D) = \emptyset$, is attached to the presented axiomatics, then A8 reduces itself to the formula: $(\forall i)(\forall X)\ [i \in T \rightarrow X \subset C_i(X)]$. Hence, the formulas: (TA), A9, A10 and A11 constitute conditions for the operator of logical consequence in the Tarskian sense.

The following taboo function can be defined:

(DF *l*) $l(D) = \emptyset$

l can be understood as a liberal taboo function, because it assigns an empty set of banned formulas to discourse D. The consequence operator indexed by this function can be called the liberal consequence operator. This operator satisfies the following conditions:

(T2) $(\forall X)[l \in T \rightarrow X \subset C_l(X)]$
(T3) $(\forall X)[l \in T \rightarrow C_l C_l(X) \subset C_l(X)]$
(T4) $(\forall X)(\forall Y)\{l \in T \rightarrow [X \subset Y \rightarrow C_l(X) \subset C_l(Y)]\}$

The liberal consequence operator behaves logically in the same way as any standard consequence operator in the Tarskian sense.
Consequence operators indexed by taboo functions form a class of etatist consequence operators when their indexes are taboo functions that take values that are not an empty set.

(DF ET) $(\forall i)(C_i \in ETAT \equiv i(D) \neq \emptyset)$

The relationships between the liberal consequence operator and etatist consequence operators are expressed in the following statements:

(T5) $(\forall i) \{C_i \in ETAT \wedge i \in T \wedge l \in T \rightarrow (\exists \alpha)(\exists X)[\alpha \in C_l(X) \wedge \sim (\alpha \in C_i(X))]\}$
(T6) $(\forall i)(\forall X)[C_i \in ETAT \wedge i \in T \wedge X \cap i(D) = \emptyset \wedge l \in T \rightarrow C_i(X) = C_l(X)]$

According to (T5), for each etatist consequence operator there are such formulas and such sets of formulas that a given formula belongs to the liberal consequence of a given set of formulas, but does not belong to the etatist consequence of the same set of formulas. (T6) states that every consequence operator acting on any set of formulas in which there are no formulas banned from the point of view of the consequence operator's taboo index, is indistinguishable from the operator of liberal consequence acting on the same set of formulas. Both statements show that etatist deduction differs from liberal deduction within a given discourse only in the range of banned formulas designated by the taboo function associated with a given consequence operator.

On the basis of A5, it can be proved that any etatist consequence operator determines the same logic (the set of logical theses) as the operator of liberal consequence.

(T7) $(\forall i)[i \in T \land l \in T \rightarrow C_i(\emptyset) = C_l(\emptyset)]$

In addition, any two consequence operators do not differ from each other in their action on an empty set:

(T8) $(\forall i)(\forall k)[i \in T \land k \in T \rightarrow C_i(\emptyset) = C_k(\emptyset)]$

Two different etatist consequence operators differ from each other, operating in the areas of banned formulas established by taboo functions constituting their indexes.

(T9) $(\forall i)(\forall k)(\forall \alpha)[i \in T \land k \in T \land \alpha \in i(D) \land \sim(\alpha \in k(D)) \rightarrow (\exists X)(\alpha \in C_k(X) \land \sim(\alpha \in C_i(X))]$

According to the hypothesis set out in the first part of the work, performative stigmatization of some sentences generated in the process of developing discourse with the property of offense, blasphemy or the prohibition activates mechanisms blocking processes of inference with the use of banned sentences. The operator of liberal consequence determines, therefore, a mental mechanism that triggers the deductive processing of discourse in situations where the participant does not recognize the premises or conclusions having the stigma of banned formulas established by any taboo function. However, when the mind captures premises and conclusions through the stigma of banned sentences established by any taboo function, then the corresponding etatist consequence operator indexed by the appropriate taboo function starts to work in the mind.

If the logic used by the participants of the discourse in its processing outside the context of sentences belonging to a particular language taboo is classical logic, then the consequence operator establishing this logic is a liberal operator. However, if the mental deduction processes carried out within a given discourse encounter "reefs" in the form of premises or conclusions belonging to a particular language taboo, then the operator of classical consequence is transformed into the appropriate operator of etatist consequence, which behaves the same as the first one in the environment of sentences not tabooed. This transformation of the classical consequence operator into etatist consequence operator is determined by the deduction structure associated with the given discourse at a particular stage of its development.

Another important consequence operator that may appear in the deduction structure of a given discourse with a language taboo is the operator of the total taboo. Its definition is as follows:

(DF t) $t(D) = D$

The following theorems characterizing the inferential properties of the consequence operator indexed by the total taboo function t can be proved:

(T10) $t \in T \rightarrow (\forall X) \; C_t(X) = \emptyset$
(T11) $t \in T \land 1 \in T \rightarrow C_t(\emptyset) = \emptyset$
(T12) $t \in T \rightarrow (\forall i)(i \in T \rightarrow C_i(\emptyset) = \emptyset)$

According to (T10), if the total taboo function belongs to the deduction structure of a given discourse with language taboo, then the set of consequences of the operator, indexed by the total taboo function, acting on any set of formulas is an empty set. On a total taboo, discourse participants can only remain silent. According to (T11) and (T12), the introduction of the total taboo function into the deduction structure of a given discourse destroys its tautological nature. This conclusion is intuitively obvious. From the point of view of the total taboo function, any statement is a breaking of the language taboo. Therefore, if experts prohibiting the formulation of any sentences within a discourse are associated with its deductive structure, then such experts invalidate the universal validity of any inferences, which consequently leads to the disappearance of tautologicity, since tautologicity is to establish logical validity seen from the point of view of each consequence operator associated with a given discourse in a given phase of its development.

3. Discourse Deduction Structures with Taboo

Different types of deduction structures with taboo can be distinguished due to their metalogical properties. In addition, one can speak of the development of a given discourse due to the transformation of its deduction structures. Thus, each discourse can be attributed to some history of its deductive transformations, distinguishing in it certain specific processes of transformation of its taboo deductive structures.

The elementary deduction structures with taboo are those that are formatted with one consequence operator and one taboo function, which is not a total taboo.

(DF. EL) $< D, CN, T> \in EL \equiv (\exists i)(CN = \{C_i\} \land T = \{i\} \land i \neq t)$

Standard elementary structures can be distinguished among the structures belonging to set EL:

(DF. ST-EL) $< D, CN, T> \in ST\text{-}EL \equiv (CN = \{C_l\} \land T = \{l\})$

Standard-elementary deduction structures with taboo are composed solely of the operator of liberal consequence and of the function of liberal taboo whose value is the empty set. Tarski's general theory of logical consequence just describes $ST\text{-}EL$ structures. The taboo in these structures is not a carrier of any "modulation" in the deduction processes implemented with the help of the operator of liberal consequence. Such standard-elementary deduction structure is associated with Peano's arithmetic.[3]

Each elementary structure of deduction develops in the process of prefabrication of a given discourse by proliferating the contents of CN and T sets. The final phase of such a process of developing a given discourse may be a situation in which the sum of the values of the family of all taboo functions is identical to the set D. These are the maximal deduction structures in the sense that any reasoning within such a discourse will appear to be prohibited from the point of view of one of the taboo functions and the corresponding operator of consequence.

(DF. MAX) $< D, CN, T> \in MAX \equiv (\forall \alpha)[\alpha \in D \rightarrow (\exists i)(i \in T \land \alpha \in i(D))]$

If a *MAX*-type structure is associated with a given discourse, anything that can be said in this discourse will offend someone (the acolyte of some taboo function). It is obvious that every deduction structure with taboo, to which the total taboo function belongs, is a structure of the type: *MAX*.

(T13) $t \in T \rightarrow <D, CN, T> \in MAX$

In the maximal structures of deduction associated with discourse D, consequence operators do not determine a set of logical theses and tautologies. The following theorem can be proved:

(T14) $<D, CN, T> \in MAX \rightarrow (\forall i)(C_i \in CN \wedge i \in T \rightarrow C_i(\emptyset) = \emptyset)$

In the discourse associated with the *MAX* deduction structure, any reasoning that is logically valid from a certain point of view is invalid from some other point of view.

(T15) $<D, CN, T> \in MAX \rightarrow (\forall i)(\forall X, a)[C_i \in CN \wedge i \in T \wedge a \in C_i(X) \rightarrow (\exists k)(C_k \in CN \wedge k \in T \wedge \sim(a \in C_k(X)))]$

Therefore, if there are inferences within a given discourse that are correct from every point of view, then such discourse is not the maximal, which means that formulas that are non-banned on the basis of any taboo function can be formulated within this discourse.

Some discourse deduction structures may have a mechanism that blocks their evolution towards achieving the maximal discourse phase. This mechanism is described by the following axiom:

(B) $(\forall i)(l \in T \wedge i \in T \wedge C_l(\emptyset) \neq \emptyset \rightarrow C_l(\emptyset) \cap i(D) = \emptyset)$

According to (B), no taboo function in the deduction structure of a given discourse stigmatizes the logical theses established by the liberal consequence operator. Thus, if the set of logical theses set by the liberal consequence operator of a given structure is not the empty set, then according to A5, each consequence operator of a given deduction structure determines a non-empty set of logical theses identical to the set of logical theses established by the operator of the liberal consequence. Thus, if there is no liberal consequence operator in the deduction structure of a given discourse, then it is impossible to introduce into this structure the mechanism described by (B) which blocks its development towards the structure of maximal deduction. For maximal deduction structures associated with the discourse at a particular stage of its development, there are no criteria for logical correctness of inference that would be jointly accepted by all elm experts. In the discourse that has reached such a phase of its deductive development, no joint discussion is possible in which representatives of each of the elm experts associated with the given discourse may participate. In such a discourse development phase, every inference raises objections from some point of view.

Any discourse in a particular phase of its development, which is associated with the structure of deduction maximal and elementary simultaneously, cannot be subject to deductive development. This kind of discourse can be called dead. It seems that this situation occurs when in a given communication community there is a strongly penalized order of silence on a given topic. In North Korea, sentences about Kim Jong-Un's disease are not spoken in public space. The operators of logical consequence constituting discourses, which are elementary and maximal simultaneously, can be described as consequence operators of silence, because they completely block deduction processes in a given discourse. Encoding them in the minds of participants in the processes of public transmission of content fulfills the function of eliminating a given domain of discourse from cultural space.[4] Empirical data, however, point to the existence of a mechanism for the elimination of silence operators from deduction structures of discourses and, consequently, to the existence of a mechanism for transforming the deduction structure which is both elementary and maximal, into a non-maximal structure.[5]

There may hold various relationships between logical consequence operators in a given deduction structure $<D, CN, T>$, such as: conflict, subordination. The two consequence operators remain in relation of conflict to each other when the product of the values of the taboo functions constituting their indexes is an empty set.

(DF C) $(\forall i, k)[\ C_i\ conflict\ C_k \equiv (i \neq l \vee k \neq l) \wedge i(D) \cap k(D) = \varnothing]$

According to (DF C), two operators C_i and C_k remain in the relation of conflict if and only if what is banned from the point of view of operator C_i is not banned from the point of view of operator C_k. The following theorem can easily be proved:

(T16) $(\forall i, k)\{i \in T \wedge k \in T \rightarrow [\,C_i\, conflict\, C_k \rightarrow (\exists \alpha, X)(\alpha \in C_i(X) \wedge \sim(\alpha \in C_k(X))) \vee (\exists \alpha, X)(\sim(\alpha \in C_i(X)) \wedge \alpha \in C_k(X))\,]\}$

According to (T16), if two consequence operators remain in the relation of conflict, there is such inference within discourse D that it is correct from the point of view of the first operator and incorrect from the point of view of the second operator, or there is such inference that is incorrect from the point of view of view of the first operator and correct from the point of view of the second operator.

There are confrontational deduction structures among deduction structures containing taboos.

(DF CONF) $<D, CN, T> \in CONF \equiv [\sim t \in T \wedge (\exists i, k)(C_i \in CN \wedge C_k \in CN \wedge i \in T \wedge k \in T \wedge i \neq k \wedge C_i\, conflict\, C_k)]$

In $CONF$ deduction structures, there are at least two consequence operators that are in conflict with each other. From (T16) follows the theorem that in the $CONF$ type deduction structure there are inferences that are correct from the point of view of one consequence operator and at the same time incorrect from the point of view of another consequence operator.

(T17) $<D, CN, T> \in CONF \rightarrow (\exists i, k)(\exists \alpha, X)\,[C_i \in CN \wedge C_k \in CN \wedge i \in T \wedge k \in T \wedge i \neq k \wedge \sim(\alpha \in C_i(X)) \wedge \alpha \in C_k(X)]$

If the deduction structure includes a liberal consequence operator and some etatist consequence operator, then this deduction structure is of the $CONF$ type.

(T18) $l \in T \wedge (\exists k)(k \in T \wedge C_k \in ETAT \wedge C_k \in CN) \rightarrow <D, CN, T> \in CONF$

In the confrontational deduction structures associated with a given discourse, there is always a dispute between competing experts in that there are inferences for the first of them that are correct from the point of view of his consequence operator and incorrect from the point of view of the consequence operator of second experts, and vice versa. Both sides of the conflict attack each other due to breaking the language taboos, because the areas of these taboos established by elm experts represented by the appropriate taboo functions are disjoint.

Between the taboo functions and respectively between the corresponding consequence operators there can hold a relation of taboo extension and respectively the relation of dominance (subordination) of one operator over another.

(DF EXT) $(\forall i, k)[\, i \text{ ext } k \equiv i(D) \subset k(D) \land i \neq k]$

The taboo function k is an extension of the taboo function i if and only if the value of function i is contained in the value of k and both functions are different. It is obvious that the total taboo function is an extension of all non-total taboo functions, and that each etatist taboo function is an extension of the liberal taboo function.

(T19) $(\forall i)(i \neq t \rightarrow i \text{ ext } t)$
(T20) $(\forall i)(i \neq l \rightarrow l \text{ ext } i)$

In discourse development practices, the taboo extension process is often started. The set of banned sentences is, for example, expanded with new sentences by introducing additional bans on speaking on specific topics within the domain of discourse. The tightening of political censorship is a paradigmatic example of this process. The final point of this process is the introduction of the total taboo function into the deduction structure of discourse in this last stage of its development, which manifests itself in the effort of political authorities to erase a given discourse from the public space of discourses[6].

The consequence operator C_i dominates the consequence operator C_k if and only if the taboo index of the first operator is an extension of the taboo index of the second operator.

(DF dom) $(\forall i, k)(C_i \text{ dom } C_k \equiv k \text{ ext } i)$

It is easy to see that every etatist consequence operator dominates the liberal consequence operator.

(T21) $(\forall i)(C_i \in ETAT \rightarrow C_i \text{ dom } C_t)$

One can distinguish the deduction structures associated with some discourses in certain development phases, in which all consequence operators are dominated by some consequence operator, which is not a consequence operator indexed by the total taboo function.

(DF DOM) $<D, CN, T> \in DOM \equiv (\exists i)[i \neq t \land i \in T \land C_i \in CN \land (\forall k)(k \in T \land C_k \in CN \land k \neq i \rightarrow C_i \text{ dom } C_k)]$

Some discourses may develop deductively in such a way that the proliferation processes of consequence operators, which generate conflicts in discourse practices, may culminate in a phase in which all etatist consequence operators are dominated by one operator. As a result of such a process, different areas of different taboos are subordinated as fragments to one language taboo correlated with the dominant operator of consequence in a given deduction structure in its specific phase of development. In other words, all sentences that are banned from different taboo points of view, at some stage in the development of the deduction structure of a given discourse, can become banned from exactly one taboo point of view.

The above-presented definitions of types of discourse deduction structures with taboo, types of taboo functions, types of consequence operators and relationships between taboo functions and between consequence operators indexed with these functions allow the construction of various idealization models of the deductive development of any discourse. In the initial phase of discourse formation, it is usually correlated with the standard, elementary structure of deduction *ST-EL*, in which elm experts do not establish any areas of discourse taboo. If, as a result of discourse development, its participants begin to produce sentences whose content somehow violates the interests of some group of discourse producers, then elm experts defending a given interest establish language taboos within the given discourse.[7] This kind of action triggers various reactions in the form of establishing other taboos. As a result of their proliferation, conflicts arise, and the space of a given discourse becomes more and more susceptible to control practices implemented from various taboo points of view. This phase of the deductive development of discourse can be called its etatization. The final moment in the development of this phase is the constitution of the maximal taboo structure of deduction. If a *MAX*-type structure is associated with a given discourse in some development phase, then the deductive processing of the given discourse is no longer controlled by tautological criteria. Then any inference within such a discourse is always invalid from the point of view of some taboos. In order for discourse to develop further, struggle mechanisms between elm experts representing specific taboo functions and corresponding logical consequence operators indexed by these functions must be activated. As a result of this struggle, the structure of discourse deduction simultaneously de-etatizes (some taboo functions and the consequence operators correlated with them are eliminated from the structure of deduction) and transforms into a structure with the dominant consequence operator. When, as a result of fights between elm experts, the function of the liberal taboo is eliminated from the deduction structure of a given discourse and, as a result of this process, the operator of the liberal consequence is deactivated, then the deduction structure of the given discourse is transformed into a slave structure because it possesses no elm experts coordinated with a liberal consequence operator who could battle all etatist taboo functions. Within such discourse, the processes of free processing of discourse sentences (content) cease to take place. It is then impossible to process such discourse only on the basis of formal and logical criteria of correctness.

The total taboo function and the corresponding consequence operator, introduced into the deduction structure of a given discourse, allow discourse annihilation. It seems that the total taboo function may appear in the deductive structure of discourse at every stage of its deductive development. The appearance of this function in the deduction structure of discourse with taboo, however, does not mean that annihilation of discourse will prove effective.

The transformation of taboo structures of deduction of a given discourse during its development is determined by out-of-logical factors. The most important of these seems to be the factor of penalty. With each deduction structure $<D, CN, T>$ there is a correlated set of penalization functions that establishes penalties of a certain intensity for breaking various taboos of discourse established by elm experts.

4. Penalty Functions in Deduction Structures of Discourses

Along with the establishment of the taboo functions, elm experts establish conventions for punishing discourse participants for committing acts of breaking language taboos. Thus, with each taboo function and the corresponding operator of consequence, the penalty function is correlated, assigning sentences, sets of sentences and inferences that break the taboo value in the form of a specific intensity of punishment. These intensities create a linear order from minor penalties to final (maximal) penalties. The latter manifest themselves by physical elimination (and even killing) of a taboo-breaking participant from the discourse. For example, for publicly calling Stalin or Hitler a criminal threatened the death penalty (shooting, sending to a gulag or to a concentration camp) in the Soviet Union and Nazi Germany. Public positive utterances on the subject of Jews during World War II were also severely penalized in almost all countries conquered by the Nazis. In the twentieth century, Khomeini imposed a fatwa on Salman Rushdi for writing the novel *Satanic Verses*. In Poland, Kazimierz Łyszczyński was killed by decapitation, with the consent of King Jan III Sobieski, for calling God a chimeric being in the treatise *De non existentia dei* [9, pp. 126-127; 5]. From the point of view of any taboo function, the intensity of penalties for breaking a taboo are differentiated on the basis of the utterance of such or other sentences or carrying out such or other inferences. It seems that the statement "John Paul II was a sinner" is penalized by Catholic elm experts with less intensity than the statement "John Paul II was a friend of pedophiles". It can be assumed for the purpose of idealization that acts of uttering sentences or making inferences that break certain language taboos are penalized with a constant intensity constituting the resultant of all the intensities of penalties imposed on participants of the discourse who break this taboo established by the given taboo function.[8]

Let *PEN* be a set of all penalty functions coordinated with corresponding taboo functions. Let p_i, p_k, ..., p_j be the variables ranging the set of penalization functions, where i, j, k represent the corresponding taboo functions. Arguments of any penalty function p_i are formulas belonging to $i(D)$, sets of formulas contained in $i(D)$, and inferences infected with the given taboo function i belonging to the set $2^D \times D$, constituted from at least one sentence belonging to $i(D)$. K is a linearly ordered set of intensities of penalties, where 0 is no penalty, and l is the maximum penalty (in the form of annihilation of a taboo breaking discourse participant). Between 0 and 1, all rational numbers are the intensities of some indirect penalties. The variables running the set of these values are: $v, v_1, ..., v_h$. The definition of set of inferences infected with the taboo function i is as follows:

(DF Infec) $(\forall X, a)[<X, a> \in Infec_i \equiv X \cap i(D) \neq \emptyset \lor a \in i(D)]$

In order for the inference to be infected with the taboo function i, the set of its arguments X must contain at least one sentence banned by this taboo function or the conclusion must belong to the set of formulas $i(D)$.

Each penalty function therefore meets the following condition:

(PEN1) $(\forall p_i)(p_i \in PEN \land i \neq l \rightarrow p_i \subset [i(D) \cup 2^{i(D)} \cup Infec_i] \times K$

The structure of the form $<D, CN, T, PEN>$ can be called the penalizing-taboo structure of deduction of discourse D. It can be assumed for the purposes of idealization that every penalty function from the structure $<D, CN, T, PEN>$ is a constant function.

(PEN2) $(\forall i, p_i)[i \in T \land p_i \in PEN \rightarrow (\forall x)(x \in i(D) \cup 2^{i(D)} \cup Infec_i \rightarrow p_i(x) = constant)]$

Since the condition (PEN1) is not specified for the liberal taboo function, an axiom can be adopted, according to which the penalty function indexed by the liberal taboo function returns a minimum value for each formula or each set of formulas or each inference.

(PEN3) $l \in T \rightarrow (\forall x) p_l(x) = 0$

Each etatist taboo function is correlated with the corresponding penalty function, which assigns their arguments a penalty value greater than 0.

(PEN4) $(\forall i, p_i)[i \in T \land i(D) \cap D \neq \emptyset \land p_i \in PEN \rightarrow (\forall x)(x \in i(D) \cup 2^{i(D)} \cup Infec_i \rightarrow p_i(x) > 0)]$

Since all penalty functions are fixed functions, one can define a function P that assigns to each taboo index the value of penalty intensity, which penalty functions, correlated with a given taboo function, assign to all their arguments.

(DF P) $(\forall i)\{i \in T \to [P(i) = v \equiv (\forall x)(x \in i(D) \cup 2^{i(D)} \cup \mathit{Infec}_i \to p_i(x) = v)]\}$

If there is a penalty function correlated with the total taboo index *t* in the penalizing-taboo structure of deduction of a given discourse, it is natural to assume that *P* function takes values from argument *t* higher than those for which *P* function takes from arguments different from *t*.

(PEN5) $t \in T \to (\forall i)(i \in T \wedge i \neq t \to P(t) > P(i))$

Due to how the *P* function works on taboo functions occurring in deduction structures of the form <*D, CN, T, PEN*>, one can distinguish their various types. In addition, as the discourse develops, the values of the *P* function from different taboo functions may change. This means that the intensity of punishment practices for breaking different language taboos in the processes of developing a given discourse may weaken or increase.

Totalitarian structures of deduction of discourses in their specific developmental phases are characterized by the fact that among the taboo functions there are those to which the function *P* assigns the maximum value (exclusion from a discourse of a participant who breaks certain language taboos).

(DF TOT) <*D, CN, T, PEN*> $\in TOT \equiv (\exists i)(i \in T \wedge P(i) = 1)$

If, in the totalitarian structure of deduction, the taboo function for which the function *P* takes the value *1*, is an extension of all taboo functions, then such a structure characterizes discourses in the development phase of the dominance of one totalitarian elm expert. It seems that the Leninist-Marxist discourse during Stalinism was in this phase. This property of totalitarian deduction structures can be described as the totalitarian monopoly of an expert institution for punishing, for example, the death of discourse participants breaking any linguistic taboos.

(DF M-TOT) <*D, CN, T, PEN*> $\in M\text{-}TOT \equiv (\exists i)[i \in T \wedge P(i) = 1 \wedge (\forall k)(k \in T \wedge k \neq i \to k \mathit{\ ext\ } i)]$

Penalties imposed on participants in the discourse may be characterized by such intensity that evokes a sense of severity. This feeling manifests itself in the state of alienation of discourse participants punished in this way for breaking a language taboo in a given discourse. The experience of such alienation causes reflexes of escape from a given space of discourse among its participants.[9] Let a be the smallest value of the intensity of the punishment causing a state of alienation from discourse. If there are elm experts in the structure of discourse who establish taboo functions that generate a relationship of conflict between the operators of consequences indexed with these taboo functions, and the function P assigns them a value of intensity of punishment causing a state of alienation from discourse, then such a structure of deduction can be called revolutionary. In such a discourse development phase, elm experts attack each other with severe punishments that cause a sense of alienation among discourse producers. In extreme cases, experts can kill each other.

(DF REV) $<D, CN, T, PEN> \in REV \equiv (\exists I, k)(i \in T \land k \in T \land C_i \in CN \land C_k \in CN \land C_i \text{ conflict } C_k \land P(i) \geq a \land P(k) \geq a)$

It seems that religious discourse during the French Revolution correlated with such a revolutionary structure of deduction. Jacobins, girondists, royalists and others killed each other in defense of their beliefs and views expressed publicly. Parties to the conflict during this revolution established their taboos in religious discourse, the breaking of which resulted in death by guillotine or assassination.[10] The intensity of the mood of the revolutionary structure of deduction, correlated with a given discourse in a particular phase of its development, increases with the proliferation of consequence operators that are in conflict with each other, and with the increase in the values of P function whose arguments are taboo functions occurring in the deductive structure of the developing discourse. The culmination of the development process of the revolutionary structure of deduction is the phase in which it takes the form of a terrorist structure of deduction. In such a structure of discourse deduction, all parties to the discourse attempt to kill each other.

(DF TERR) $<D, CN, T, PEN> \in TERR \equiv <D, CN, T, PEN> \in REV \wedge (\forall i, k)(i \neq k \wedge i \in T \wedge k \in T \wedge C_i \in CN \wedge C_k \in CN \rightarrow C_i$ conflict $C_k \wedge P(i) = 1 \wedge P(k) = 1)$

Some specific processes of discourse development can be distinguished:

(i) In the initial phase, the standard elementary structure ST-EL correlates with the discourse. The deduction processes in this phase are governed by some operator of liberal consistency, defining specific logic (in particular, classical logic). Then, the ST-EL structure, which is a fragment of the structure $<D, CN, T, PEN>$, where $PEN = \{p_l\}$ and, as a consequence, $P(l) = 0$, undergoes proliferation processes, as a result of which subsequent deduction structures with etatist operators appear. Along with the constitution of such deduction structures in the space of a given discourse, elm experts assign, by virtue of the P function, taboos functions in these structures to values less than a. These processes lead to the constitution of the $CONF$ type deduction structures associated with the given discourse. Disputes and conflicts within the D discourse cause, as a result of a process of escalation, the transformation of penalizing taboo structures of deduction into revolutionary type structures REV, which can transform into $TERR$ type structures. The final process is the appearance in the space of a given discourse of totalitarian monopolistic structures M-TOT. The transformation of $TERR$-type structures into M-TOT-type structures is a characteristic feature of the discourse development phase, which can be described as its terrorization. A good example of this process is the situation in Cambodia during the reign of Pol Pot. Any deduction regarding politics, social or religious matters was banned, and breaking the bans resulted in death.

(ii) When the discourse finds itself in a phase in which it is associated with some type of MAX deduction structure, it is susceptible to processes of de-etatization, i.e. reduction of etatist consequence operators within such a structure. Along with this process of de-etatization, depenalization processes may take place, i.e. decreasing the value of the P function of the arguments that are taboo functions. It is not uncommon to see the disappearance of etatist consequence operators indexed by taboo functions to which the P function assigns penalty intensity values close to zero in the structure of a given discourse at a particular stage of its development. The culmination of such a process is the constitution of a standard, elementary structure of deduction for a given discourse.

(iii) Some discourses from the initial phase, when the *ST-EL* deduction structure is correlated with them, develop so that their initial deduction structure transforms into an *M-TOT* structure with exactly one taboo function, which is a total taboo function. When the standard, elementary deduction structure of discourse in its initial phase transforms into the structure $<D, CN, T, PEN>$, where $T = \{t\}$ and $P(t) = 1$, it means that elm experts attempt to annihilate a given discourse in its bud (due to the extreme threat to their interest caused by the development of a given discourse).

The sketched theory allows for formal modeling of various discourse development processes. However, it needs its supplement in the form of a theory describing the functions of mutual transformation of penalizing taboo deduction structures.

5. Final Notes

The above-presented theory of consequence operators indexed by taboo functions requires its development towards the theory of transformation of deduction structures that are associated with discourse during its development. Within the discourse, various narratives are created on a given topic. Through semantic relations, they are tools for creating various narrative worlds (mental worlds). The deduction processes implemented by discourse participants are not only the processes of transforming discourse sentences. They are also processes of transforming various contents in the narrative worlds of a given discourse. Elm experts who impose taboos on discourse establish sets of sentences banned in various narrative worlds of the space of a given discourse. At the same time, through semantic relations, they point to those fragments of these worlds that for some pragmatic reasons (interests) should not be developed in the processes of their prefabrication or even should disappear from them. The transformation of discourse deduction structures is the process of transforming the logical architecture of the discourse space into another architecture. The theory of such transformations will be a description of just such possible logical and architectonic changes of the structures of discourse space.

The scope of application of the presented theory is wide. The central field of theory application are the processes of transformation of ideological, political, religious and even legal discourses. For cases of such discourses (ideological struggles at the beginning of Christianity, Cathar genocide, fascism in the humanist discourse, communist discourse) the analytical application of the theory is seen as obvious. Such examples can undoubtedly be multiplied. The presented theory can also be used in the analysis of the history of scientific narratives. Its conceptual tools could be used in research on scientific revolutions. All these applications would reveal a new field of research. In the theory of discourse, first of all, attempts are made to explain how the content, grammatical forms and illocutionary forces of speech acts influence the phenomenon of power in political, social, gender and other perspectives. Discourse researchers, however, do not notice the fact that the styles of logical processing of these contents are also a factor influencing the production of discourse for various interests in manifesting power and forcing obedience.

Acknowledgement

The article was written as part of project No. 2016/21 / B / HS1 / 00821 financed by the National Science Center. The source of inspiration for the presented concept is the book by A. Schumann *Talmudic Logic*, in which the author (Jan Woleński's follower) formulates a postulate to examine the impact of methods of logical information processing on totalitarian social practices. He states: '/…/ in the totalitarian thought we detect the priority of confirmation (modus ponens) and deductive syllogism, this priority undoubtedly speaks of appropriate prereflexive presuppositions. Nonrandomly, in totalitarian cultures schemata of confirmation and deductive syllogisms are preferred, because in these cultures thinking is totalitarian and holistic. For instance, in the Soviet Union of Stalin's epoch the following modus ponens was often used: /…/' [8, p. 157]. Schumann's arguments in favor of the proposed research paradigm are very weak. However, the postulate to study logical mechanisms of information processing in the context of generating totalitarian practices should be considered innovative. In the light of this approach, the mental structures responsible for our totalitarian thinking are logical. Schuman believes that mental deduction processes, governed by classical logic, generate totalitarian attitudes. The conclusion of my work is antagonistic to Schumann's position. The classical operator of logical consequence turns out to be in its character a liberal operator, not an etatist one. Etatism is understood in the article as a category referring to an ideological attitude that generates meta-behaviors aimed at controlling and eliminating other behaviors from public space that violate some vision of the world. According to this explication, etatism is the opposite of liberalism understood as an attitude of refraining from controlling and eliminating other behaviors from public space.

References

1. Fodor J. A. *Eksperci od wiązów. Język myśleński i jego semantyka*, transl. by M. Gokieli, Warszawa: Fundacja Aletheia, 2001.
2. Hałkowska, K. *Algebry związane z teoriami* zawierającymi *definicje warunkowe*, Warszawa, Wrocław: Państwowe Wydawnictwo Naukowe, 1979.
3. Kelly, J. N. D. *Początki doktryny chrześcijańskiej*, transl. by J. Mrukówna, Warszawa: Instytut Wydawniczy Pax, 1988.
4. Nowak L. *U podstaw teorii socjalizmu. Tom I. Własność i władza*, Poznań: Wydawnictwo Nakom, 1991.
5. Nowicki, A. R. *Kazimierz Łyszczyński 1634 – 1689 [1]*, 2007, retrieved from: http://www.racjonalista.pl/kk.php/s,5202/q,Kazimierz.Lyszczynski.1634.1689
6. Putnam, H. Znaczenie wyrazu "znaczenie", In *Wiele twarzy realizmu i inne eseje,* transl. by A. Grobler, Warszawa: Wydawnictwo Naukowe PWN, 1988, pp. 93-184.
7. Sawicka, G. Konwencja a tabu językowe, *Język a Kultura* 21, 2009, pp. 31-46.
8. Schumann, A. *Talmudic Logic*, London: Individual author and College Publications, 2012.
9. Skoczyński, J., and J. Woleński. *Historia filozofii polskiej*, Kraków: Wydawnictwo WAM, 2010.
10. Zgółka, T. Retoryka tabuizacji, *Język a Kultura* 21, 2009, pp. 23-29.

Notes

1. Language taboo is the subject of linguistic and ethnolinguistic research. Researchers distinguish language taboos from cultural

taboos. The latter are understood as a set of socially established prohibitions on certain actions in relation to specific objects, situations or facts. They can manifest themselves in cultural spaces of various types: religious, magical or political [7, pp. 31-34]. Language taboos, however, are usually understood as a set of prohibitions on the use of certain expressions and on speaking on specific topics in a given community [10, pp. 24-25]. The violation of prohibitions that make up language taboos, as in the case of cultural taboos, is punished with various penalties.

2. Inspired by Putnam's concept, Fodor introduced the notion of experts to the language of semantics. According to Fodor, experts are the guardians of meanings of terms by setting the conditions for the truth of thoughts expressed with their help [1, pp. 33-39). According to Putnam, there are experts in every language community who know the meaning of certain terms, so that other language users can use them efficiently without knowing the meaning of these terms [6, pp. 112-115]. I will refer to Fodor's experts as elm experts in this paper. This concept can be extended by giving them an additional role, namely, setting logical inference norms and hermeneutic norms for a given discourse along with establishing a specific language taboo and rules for penalizing taboo breaking practices. The guards of the Soviet revolution, namely NKVD officers and members of the central committee of the Bolshevik party, are a good example of elm experts. Lenin called them the vanguard of the proletariat, devoid of the so-called false consciousness. Another equally good example of elm experts are the Guardians of the Iranian Revolution. The intellectual leaders of various ideological movements, often referred to by their followers as gurus, are actually fulfilling the missions of elm experts within their discourses. Popes, prophets, missionaries, holly-men and sorcerers typically function in their ideological communities as elm experts setting up various taboos.

3. It seems that in relation to arithmetic theories regarding numbers other than natural numbers, e.g. rational, real or even imaginary numbers, one can speak of a language taboo. In the languages of such theories, grammatically correct formulas devoid of mathematical meaning can be constructed. For example, in rational number arithmetic, the formula: $1/0 = 0$ is not false, but rather devoid of arithmetic sense because there are no fractions whose denominator is the number 0. In various arithmetic theories, the so-called indicators

of meaningfulness of defined formulas are given in conditional definitions. For example, in the definition of decimal logarithm such a clause is used. It is the formula: x> 0. The definition takes the shape: $(\forall x)[x > 0 \rightarrow (y = \log(x) \equiv 10^y = x)]$. Although the expression "y = log(-6)" is correct from the point of view of the syntax of the real numbers arithmetic, it is meaningless. Such formulas may just be tabooed. Some logicians try to show that mathematical deduction realized in the environment of such formulas must be based on an adequate logic of nonsense [2].

4. Pedophilia among Catholic priests or the financial activities of Saint Mother Teresa of Calcutta for many years were subjects to the so-called conspiracy of silence in the cultural space.

5. The topic, which was silent in public space at the price of losing life, often returns after some time to the public agenda. Stalin's crimes were the subject of silence during his reign. Khrushchev broke this collusion of silence with his famous paper during the 20th Congress of the soviet communist party.

6. After Germany invaded the Soviet Union during World War II, the discourse on the Molotov-Ribbentrop Pact was subjected to such an operation. Expressing any sentences on this subject was prohibited in the USSR and threatened with penalty in the form of the death sentence or exile to the Gulag. Similar practices were initiated in relation to the Katyń discourse in Poland during the Stalinist period.

7. For example, in the early stages of the formation of the Christian discourse, various doctrines appeared that were stigmatized with the marker of heresy by some producers of this discourse. A model example is the doctrine of Arius, according to which Jesus Christ is not God the Father. In the 4th century, "Nice elm experts" condemned Arianism for questioning the dogma of the Trinity. In this way, a taboo was established, breaking of which resulted in being burned at the stake several hundred years later. Questioning the dogma of the Trinity harmed the interests of Christian hierarchs advocating the unity of the Roman Empire (on disputes with Arianism within the early Christian discourse, see [3, pp. 171-190]).

8. Breaking Islamic taboos today is punished more heavily than breaking Catholic taboos. Participants of religious discourse who break Islamic taboos are most often threatened with killing, as evidenced by the massacre at the editorial staff of the satirical weekly Charlie Hebdo in Paris in 2015. The punishment for participants of

religious discourse for breaking Catholic taboos are usually public stigmatization of such people, carried out by Catholic elm experts. The death penalty for questioning Christ's sanctity or for caricaturizing him is absent currently, whereas attempts to kill infidels for their blasphemy against Allah are a systematic phenomenon.

9. The notion of alienation of a discourse participant should be understood similarly to the category of alienation of labor in L. Nowak's philosophy of non-Marxian historical materialism. According to this philosopher, there is a certain value of the level of alienation of labor (called the value of outclassing) at which the ability of direct producers to resist the owners of means of production disappears. A similar situation can be found in the case of activity in the field of discourse production. The imposing of severe punishments for breaking language taboo by the elm experts on non-expert discourse participants leads to escapist actions in relation to a given discourse among the punished, and for retaliation among experts remaining in conflict with the former (see on the topic of labor alienation, [4, pp. 31-33])

10. The advocate of the revolution, Jean-Paul Marat was stabbed by an adversary of violence, while King Louis XVI was guillotined. Marat demanded death for the king.

The Normative Permission and Legal Utterances

Marek Zirk-Sadowski

University of Łódź
Kopcińskiego 8/12 Street
90-232 Łódź, Poland
e-mail: msadowski@wpia.uni.lodz.pl

1. Introduction

In the current theory of law practically only the contention regarding a logical status of rights has managed to reveal the whole complexity of this notion. We omit here the question of the so-called logic of norms, believing that for reason of the assumption of anticognitivism, more convenient could be the consideration of legal inferences within the language of deontic logic. It shall be then remembered, as in essence a discourse on the topic of the logical status of rights related to deontic propositions stating that something is permitted. What is conspicuous here are some gaps within the pragmatics of such logic systems. For this reason, their authors restrict the area of their studies only to the statement that using deontic propositions is always relativized to some normative system, but not the whole deontic logic within which we use them [6, p. 117].

With rare exceptions those logic systems use three types of operators: injunction, prohibition and permission. These three operators are a reflection of influences exerted on deontic logic systems modal logic. Usually a necessity was associated with an injunction, impossibility with prohibition and a possibility (or permission) with a right. There are however known systems based on analyses carried out by W. N. Hochfeld which within the deontic logic use a much greater number of operators in the process of translating the content of norms into deontic propositions.

The first remark then, suggested by logical analyses of a right, indicates a fact that the problem of rights appears markedly only in deontic logic systems, which use three operators. We could include within these logic systems also those introducing a fourth operator of an indifferent operation, when the indifferent preceded by the external negation gives a formula: ~IPdi OP or Fp.

Secondly, an entitlement within these logic systems is treated as the so-called weak permission or the so-called strong permission. The contention on the topic of the status of the logical permission relates mainly to the strong permissions. Regarding the weak permission we could say that it is framed as a non-prohibition. The permission framed as non-injunction would have to include what is prohibited. It appears as an intuitive statement that permission equals a non-prohibition, because it includes the mandatory dimension or the indifferent [6, p. 116]. Within this framework a statement of a type: "I could do what they recommend me to do (what I am commanded to do)" is treated as intuitive, which perhaps should be understood as "I have an entitlement to do, what they recommend me to do (what I am commanded to do)".

A question arises as to whether such a framed entitlement does not rather relate to intuitions associated with the notion of competence. Also, a statement that an entitlement includes the dimension of the indifferent is not free from doubts. I have to take as intuitive (obvious) an expression: "I could do what is legally neutral". A permission for such a formula requires however an acceptance of a certain concept of the qualification-driven completeness of the legal system. The permission as a non-prohibition brings then to existence also some interpretative problems. It appears that it has to assume the above demonstrated assumptions.

Logicians accept that a weak permission, or "p is not prohibited", could be stated without referring to the function of behavior control. What suffices is the analysis of the dimension of what is ordered and forbidden. In other words, to state that I am entitled to a certain behavior, in the sense of a weak entitlement what is sufficient for me is a description of the dimension regulated with a prohibition and an injunction.

While the dimension of the weak permission could be relativized till the moment of obtaining the division into the dimension of the injunction and the prohibition, it is not so obvious in the case of the so-called strong permission. Arguments appearing in the literature on behalf of the separation of the strong permission are as follows:

1) the existence of the autonomous elements of a normative system is recognized, regarding different injunctions and prohibitions; an often currently provided example consists of the so-called secondary rules within the construction of law presented by H. L. A. Hart;

2) lawyers experience an intuition that the concepts of "being entitled", or "has a right to" express a particular normative content, which is richer than the content of the non-prohibition.

The arguments provided made the logicians seek a deontic functor which would allow for respecting these two arguments. Let us notice that they do not have the same persuasive power. The first one refers to some assumed construction of the legal system. The deontic logic would then have to be based on the relativization to some theoretical notion of the legal system. The second argument is at least of a linguistic nature, if not even of the philosophical nature. This is because it states that the linguistic function of norms includes something more than a prohibition and an injunction and that it includes a strong permission, expressed in the language with the utterance "he/she/it has a right". The first argument follows then a reconstructive attitude towards the legal language, and the second one the descriptivist attitude.

If a logician would like to account for those wants and build a deontic logic including strong permissions, he or she has to accept two assumptions, as was shown by G. H. von Wrighte. First, it is necessary to tell the difference between a normative dimension and the out-of-norms (indifferent) dimension. Second, one would need to assume a possibility of formulating deontic propositions relativized to specific norms of the strong permissive nature.

Based on these assumptions, K. Opałek and J. Woleński attempted to demonstrate that elimination of strong permissions is however necessary [6, p. 121].

Consistently with the above mentioned assumptions they distinguish four dimensions of applying norms, normalizing (or regulating): O (an injunction), Ps (a strong permission), F (prohibition), I (the indifference), of which O, Ps, F belong to the normative dimension, and I to the out-of-norms dimension.

Because within the deontic logic we usually accept a proposition that Op leads to Psp, or the dimension of Ps includes an injunction and "something more", it is an interesting question to pose, as to what dimension ~p belongs, if p is strongly permitted. The authors answer this question in the following way:
If ~p belonged to O, then p would have to belong to F;
If ~p belonged to F, then p would have to belong to O and then consistently with the thesis that Op leads to Psp, p would be Ps. It would introduce an interpretative paradox, that one needs to know Op, to state that Psp, If ~p would belong to I, then p would also have to belong to I.

If so, then only accepting that ~p also belong to Ps could guarantee a separation of Ps. Accepting such a thesis and a statement that Op is a necessary condition of Ps. In relation to the above the authors propose a distinction: between "Ps sensu stricto", which includes the marked dimension (?) and "Ps sensu largo", which includes (O) and (?).

Based on this distinction, they propose a thesis that Ps sensu stricto is an analogon of the indifference (strong indifference), while Ps sensu largo is an analogon of the weak permission. It then leads to the rejection of the thesis that rights are norms. It is a consequence of the accepted by the cited authors conception of the function of behavior control. It follows from their paper that they understand in this way a capacity for expressing autonomous generating of the division of the universe of normalizing [6, p. 124].

The analysis of the logical status of rights or entitlements leads us then to a thesis that within the deontic logic systems using three operators, the question of the entitlements could be reduced to the way of comprehending the function of the act of speech, and particularly the function of behavior control. The logical level of the here presented analysis turns out then to be coming from the relations to linguistic findings, and more precisely speaking – the sociolinguistic level findings. Two competing theses emerge on that level. The first one states: the concept of the function of behavior control does not allow for including the entitlements or rights within the category of norms. Taking such a thesis as valid excludes entitlements or rights from the range of the deontic logic.

The last of the mentioned theses provides us with two possibilities of resolving the question of entitlements or rights: a) delete the entitlements from any interests of lawyers, b) recognize that these are expressions, in which there is a different function (e.g. performative function) built upon the function of behavior control, and only the analysis of the whole multi-layer act of communication could allow for distinguishing speech acts known as entitlements or rights. As it seems, to the authors of an article titled: On the disagreements regarding the so called 'permissive norms' [7, pp. 57-64].

Making a choice for one of the above-mentioned theses whose catalogue could be found as insufficient, depends on the accepted socio-linguistic assumptions. The problematic matter is that empirical studies are highly difficult to carry out, and above that, they have to be preceded with some determinations of a philosophical nature.

Within the currently demonstrated discourse what is conspicuous is the acceptance of the opponents of the thesis of an assumption that there exist permissive norms, as to that it is possible to differentiate entitlements only at the level of legal regulations. In other words, an expression "he/she/it has a right" is often an indispensable element of the conventional activity of constituting a legal regulation. This means that a legislator is not able to express with an exact list of injunctions and prohibitions their normative preferences. For instance, in order to protect interests of an owner the legislator would have to formulate a huge number of injunctions and prohibitions, which would include all the possible violations of the law of ownership. Because it is technically impossible, the legislator applies a facilitation providing an owner with a right or an entitlement to use a possession. It is possible only at the level of the legal regulation; if we would like to formulate a norm based on these regulations, then it could be only an injunction or a prohibition, because only they unambiguously determine our way of acting, and so they fulfill the function of behavior control, which is practically a norm. The opponents of the normative characteristics of entitlements or rights simultaneously accept then two theses: 1) an entitlement is logically related to an obligation; 2) permissions are pragmatically indispensable for transferring normative preferences [8, p. 64]. The price for this is the above-mentioned linguistic construction, in which a right or an entitlement is not related to a function of behavior control, but a performative function. This follows the fact that the entitlement is only the element of activity of conventionally constituting a legal regulation [7, p. 60].

It appears then that the essence of the matter of contention once again lies within the concept of the linguistic function and specifically it depends on what is understood by the function of behavior control. Accepting a thesis that the essence of the function is the fact of autonomously generating the universe of normalizing, we drastically narrow down the concept of a norm. What strikes us here is also a certain inconsequence. An injunction also does not direct in a certain sense someone's behavior, it only says what must not be done, without positively outlining the behavior.

Determining in such a way the borders of the concept of a norm, we take a reconstructivist stance towards the language. It should be remembered however, that as we have shown, a reconstructionist may also demand to recognize normative characteristics of permissions, for instance referring to the assumed concept of the legal system.

2. Pragmatic Concept of Legal Norms

Within the further part of the article we shall try to point out at least a part of the problems, which are brought to life with the pragmatic concept of legal norms. Only solving them will enable us in future to conclude the dilemmas which are created by the so-called permissive norms. As it seems, such questions are revealed by an analysis of the mere pragmatics, and later a concept of the function of behavior control built upon it.

The theory of pragmatics is a part of a general sign theory. It is possible then to talk in relation to it about internal and external effects of accepting a determination of pragmatics. The internal questions relate to associations of the theory of pragmatics with the remaining parts of semiotics, i.e. to syntax and semantics. The external problems relate to associations of pragmatics with other scientific disciplines, which are not included in semiotics [5, pp. 217-245]. Because within the theory of law – as in the case of permitting norms described above – internal problems of pragmatics are usually studied, we shall then try to limit our considerations in the same way.

Within the logical approach to pragmatic problems we are more interested in the question of the pragmatic language than the question of the mere theory of pragmatics. This significant distinction is useful in understanding the way in which logicians deal with the problem of pragmatics. Within a logic system the notion of pragmatics, or a part of pragmatic language, is related to the requirement of adjusting formal languages to deictic features of natural language. While then within theories of pragmatics – such as the above analyzed theory of Ch. Morris – pragmatics is directly the object of studies, it only has to be reconstructed from pragmatic languages.

The theories built of pragmatics show a tendency to introduce notions taken from behaviorism. Usually they are then some versions of the behavior theory [4, p. 60]. While in studying pragmatic languages built by logicians we state that the here occurring concept of pragmatics is applicable only in analyzing artificial languages, which could be interpreted in small pieces of the natural language. Theories of pragmatics and pragmatic languages use the term "pragmatics" for completely different purposes and only a very general theory of language could combine them. Modern semiotics could not fulfill this task for reason of the lower extent of formalization which it reveals. Therefore, proving a pragmatic equivalence between normalizing including permissions and a given set of injunctions and prohibitions is a doubtful matter. For now such equivalence needs to be a priori assumed.

Lack of an exact determination of pragmatics is the main reason of gaps within the construction of the linguistic function. Such inaccuracies cause a series of astonishing philosophical consequences built from the perspective of pragmatic notion of a legal norm. We shall try to demonstrate them within the next point.

The basic feature of this approach is taking a legal norm as a result of a conventional activity. There is not room to go into detailed analysis of the pragmatic criteria of separation of the expressions of legal language. It is sufficient to state that, at least in the Polish theory of law, they come near to the model of communication suggested by R. Jakobson. It is perhaps also linked with British analytical philosophy, the works of J.L. Austin and Nowell-Smith, i.e. the trend of the so-called multi-functionalists.

Jakobson's model starts from four basis concepts: the act of communication, the context, the function of language, and the concept of the dominating function derived from the latter.

In spite of the evident advantages of this model it causes a number of troubles which are difficult to omit by theoretical manipulations. Treating a legal norm as the effect of a conventional activity is the basic feature of this approach.

We could say that an utterance "a legal norm exists" is equivalent to a statement that by constituting that norm rules of cultural interpretation have been fulfilled, characteristic for the symbolic activity of constituting a legal norm. In other words, the nature of a legal norm is not fully explained with the analysis of the function of behavior control, because its fulfillment decides merely about the normative character of an utterance. A statement that a given norm is a legal norm, requires an analysis of this norm from the perspective of the performative function. A similar statement could be referred to the axiological questions, except for a difference that instead of the function of behavior control we should talk about the expressive function. This type of framing the problem of norms assumes however a normative theory of culture. This means that the culture is created by the total of patterns of behaviors, which was created and perpetuated within the process of social interaction occurring within a given community.

If we realize such nature of the assumed notion of a culture, then an analysis of legal norms reveals certain concerning features. The basic philosophical problem, which was brought to existence by the semantic theory of a norm, was a danger of falling into such philosophical solutions which at the ontological level or epistemological require negating the opposition between the Is-Ought.

The pragmatic concept of a norm allows for omitting this problem and limiting considerations only to the level of language. Accepting one or a different ontology seems to be neutral for this concept. The semantic theory of a norm requires introducing a concept of a meaning, and so referring to the category of being, if the term of a "meaning" is not used with a persuasive sense only. For it happens that a term of a "meaning" is also used in pragmatic considerations, which has that effect that it is necessary to construct two notions of a meaning. The first one which relates to the semantic relation and the second, which in essence means some feature of a whole act of communication, allowing for identifying an act of communication. A good illustration of this question is the mentioned concept of a permission or a right, in which we could recognize a permission to be a result of an effect of performative function, upon which other elements of an act of communication are built, whose essence is issuing a legal regulation. It is then not about a known distinction between a sense and a meaning, because both these terms refer to a semantic relation, but as if about a certain ideal sense, which unifies an act of communication.

Excluding such persuasive use of a term of a "meaning", applying semantic theory of a norm is related to the risk of disrespecting the difference, which has place at the ontological, epistemological and linguistic level between the Is and the Ought. An advantage of pragmatic construction is to rely on the complete bypassing these issues. Meanwhile, if we look more closely at the construction of rules of cultural interpretation, which decides about our calling some utterance a legal norm, then it is easy to state that the mentioned opposition is not omitted at all, but it is transited to the higher level. Instead of contemplating the problem of obligation at the level of an utterance, it is transferred to analyses related to the concept of a culture. The opposition of a being and – an obligation becomes replaced with the opposition of a culture and nature. Only at this level the mentioned philosophical problem could be resolved. A question then arises as to whether it is then a theoretic-legal problem. It seems that it is rather related to the philosophy of culture. Removing philosophy even from such understood pragmatic theory of a norm is however impossible. Let us try then to consider where the most difficult problem could be found.

We have talked about a double use of the term of a "meaning". If we reject such erroneous theoretic intervention, we will not remove the problem which arises on this occasion. Let us call this question a problem of identifying an act of communication. It appears when we decide to prefer the multifunctionality of an act of communication. Usually it is assumed that an act of communication is able to simultaneously perform several functions. The context in which an act of communication appears decides which of them is the dominant. Depending on the one which dominates within a given context a type of a formulated utterance is determined: a descriptive proposition, a norm, an assessment, etc.

Domination of a function within a given context does not mean however, that the remaining functions disappear. Even if we introduce a distinction of the actual and a potential fulfillment of a function, then still a question arises about how an "identity" of a given act of communication could be found, due to which we would be able to state that we are still considering the same act of communication. It is also from a different point of view a significant question. Within this framework a border between the language and a situation in which it is used is blurring. The language or rather speech appears to be the whole event, while the mere utterance just one of the elements of that event. The question about the potential of identifying an act of communication is then also for this reason significant.

Solving this problem appears to be possible in a few ways. The first depends on recognizing that what sustains an act of communication as a certain entity is a reference function. It somehow constitutes the deepest layer of a studied act, what seems to be a close approximation of that solution are suggestions of Z. Ziembinski [9, p. 115]. In the case of normative utterances difficulties emerge, when we ask how the reference function could be fulfilled with the functors of the "I ought to" type. An example is the considered in the previous points problem of an entitlement or a right.

We could obviously assume that they never occur independently, but as parts of an utterance and a notion of fulfilling a function should be referred to the intention of the whole utterance. It is however always necessary to resolve within a proposed solution the problem of criteria of distinguishing a statement of something from describing something. This difficulty we try to omit with a different construction, in which instead of recognizing the reference function as an elementary in some way, an additional notion of a "plot of the utterance" or "a propositional act" is introduced [2, pp. 84-97]. Their theoretic role is interrelated, but because a notion of a propositional act is more precise, we shall try to talk about the currently interesting for us problem by using it.

A propositional act is an act of speech considered only as an indicator of some object and a pronouncement or a judgment regarding relating to some of its features [2, p. 86]. We are not that much interested in the way of pronouncing that feature of an object while studying it. A hypothetical situation of a lack of a propositional act within an act of speech would cause for instance that in expressing a question we would have to separately inform the listener about the content of that question. Appropriately this problem appears within normative utterances. All the functions of an utterance would be therefore superstructured with the so understood propositional act. Consistently with the intention of the cited author it should rather be said that it is about dividing a propositional act and an appropriate illocutionary act. A construction of a propositional act seems to be more persuasive than the earlier proposed solution. Basing the whole act of communication on the reference function (a semantic relation) means in fact a return to the old problem of differentiating a being and an obligation, although, to omit it, a pragmatic analysis was introduced. The notion of a propositional act could not be confronted with this objection, but it raises others, not less complex philosophical problems.

A question arises as to whether an act of speech could be considered a propositional act whether it does not require introducing certain epistemological assumptions. For this act is not an ostensive definition despite that its part is indicating a certain object. A propositional act could refer both to perceptual objects, as well as theoretic, while the ostensive definition is related to perceptual terms.

What is then the mentioned 'indication' within determining a propositional act?

Similar questions are posed by 'stating a certain feature'. Consistently with assumptions introduced by the cited author, it is neither a statement nor a description which is a kind of a statement. Where do then features of an object come from and what is the nature of pronouncing them? We could try to assume that it is a certain theoretical construct, the constantly occurring factor or a category of an act of communication. For the author of this construct it is however just an act of speech, and an act of speech is a case of using an utterance. The essence of the problem is included then within the capacity of indicating an object and pronouncing its certain feature in relation to it in a neutral way, i.e. without recognizing the way of pronouncing or judging. The question about the conditions which have to be met to make this stance possible is at the same time a question about philosophical consequences of the construct of a propositional act.

3. Jacobson's and Husserl's Approaches

It appears as justified a statement that the full explanation of this type of a construct we could find only within phenomenological contemplations From the theoretic point of view, we could talk about relations of a construct of an act of communication in Jacobson's and Husserl's works. A construct of a pure sense to which a notion of a propositional act is similar is possible to justify only with acceptance of transcendent reductions. A transcendent reduction is a fulfillment of a stance of suspending everything that is external, substantial and revealing or uncovering what is immanent within consciousness. Externality is not only empirical reality within the traditional meaning of this word, it is also every real being and so also the psychic being. The starting point within such limited considerations is – according to E. Husserl – consciousness in the sense of Cartesian cogito, understood as any experience of own 'I' within its particular shapes, how: I perceive, I feel, I desire, etc. They are contemplated as a stream of experiences. Every such cogitatio is intentional and it possesses its own essence, which should be captured within its particularity [3, p. 99]. The notion of a propositional act or a plot of an expression could be interpreted as intentionality of an act of speech. More visible becomes this analogy, if we use the developed concept of intentionality, i.e. when we introduce a distinction of two dimensions of an intentional experience: noesis and noema [10, p. 93].

Currently we could say that the construction of a plot of an utterance and of a propositional act could be ordered and assigned to Husserl's notion of noema, the pure sense of an object. If so, then from the pragmatic concept of an utterance, and so also the pragmatic theory of norms, it is possible to derive assumptions whose content appears to be consistent with propositions of E. Husserl's phenomenology. An attempt at escaping philosophical problems, characteristic for the semantic concept of a norm, leads to new philosophical questions which could be linked to concepts assuming the so-called pure consciousness or, in other words, the ideal sense.

Limiting a notion of the function of behavior control to the extent of independently generating the universe of normalizing, though it is easy to carry out at the logical analysis level, it appears as risky from other points of view. If we would like the deontic logic which we construct to account for the fact that a permission is pragmatically indispensable for transferring by the employer their normative preferences, we have to resolve the consequences of the accepted function of behavior control, which arise at the socio-linguistic level. Meanwhile it turns out that for reason of the lack of a useful for a lawyer theory of pragmatics, there is no evidence that pragmatically speaking a strong entitlement or a strong right could be expressed with a list of injunctions and prohibitions. Apart from that limiting permissions only to the language of regulations of the law makes one accept a construction of an act of communication which could find the full justification only within Husserl's construction of the noema.

The normative permission is a good example of difficulties which are brought to existence by pragmatic concept of a norm. It appears that a theory based on the concept of an act of speech fulfilling multiple functions must solve a series of new philosophical questions. The accepted convention of the function of the language has consequences within both a logical analysis of a norm, as well as in studies about conventional activities of law making.

These problems reveal a characteristic way of using logic, which can be called a weak phenomenology. The analytical attitude and the phenomenological attitude are often complementary. However, the phenomenological attitude is not revealed by the lawyers. In particular, the pragmatism of norms hides its phenomenological nature, giving the appearance of a direct knowledge of normativeness.

Contrary to appearances, a weak version of phenomenology is not a direct cognition. It only serves to expose the language of the subject, but the analysis is not intended to get to the reality behind the first language. The purpose of the analysis is to translate these expressions into simpler expressions, but already in metalanguage.

Condemnation to the representationalism makes it impossible to achieve the transparency of the sign and an intentional reference to reality [see 1, ch. 6]. Meanwhile, these are two goals to which Husserl's phenomenology leads. The reversal of the direction of analysis towards the construction of metalanguage as an end in itself causes the original function of cognition, i.e. reaching reality through natural language, to be disrupted. The only goal of the analysis is to transfer the language to higher levels. In this way, the reality built by analysis is created but not recognized.

The direction should be the opposite, that is, revealing reality, striving for clarity of the first objective language and already directly anchoring it in reality. This is especially important for the practical sciences, which undoubtedly include jurisprudence.

References

1. Dębowski, J. *Bezpośredniość poznania. Spory – dyskusje – wyniki*, Uniwersytet Marii Curie-Skłodowskiej, Wydział Filozofii i Socjologii, Lublin: 2000.
2. Gizbert-Studnicki, T. Stwierdzenie jako akt mowy, *Studia Filozoficzne* 3, 1973, pp. 84-97.

3. Husserl, E. Idee I, Warszawa: PWN, 1975.
4. Kutschera von, F. *Philosophy of Language*, Dordrecht-Holland/Boston-USZA, Reidel Publishing Company, 1975, Synthese Library, vol. 71.
5. Lieb, H. H. Pragmatics and non-semiotic disciplines, In Kasher A. (eds) Language in focus, Dordrecht-Boston: 1976, Synthese Library, vol. 89, pp. 217-245.
6. Opałek, K., and J. Woleński. On the so called weak and strong permissions, *Philosophical Studies* 8, 1974, pp. 115-124.
7. Świrydowicz, K., S. Wronkowska, M. Zieliński, and Z. Ziemski. O nieporozumieniach dotyczących tzw. 'norm zezwalających', *Państwo i Prawo* 7, 1975, pp. 56-62.
8. Woleński, J. Przyczynek do analizy dozwolenia, *Państwo i Prawo* 1-2, 1982, pp. 61-64.
9. Ziembiński, Z. Podstawowe problemy prawoznawstwa, PWN, Warszawa 1980.
10. Zirk-Sadowski, M. Powinność i wartość w fenomenologicznej filozofii prawa, *Zeszyty Naukowe UŁ* 28, 1978, pp. 21-34.

www.ingramcontent.com/pod-product-compliance
Lightning Source LLC
Chambersburg PA
CBHW071652160426
43195CB00012B/1445